# Lecture Notes in Control and Information Sciences

Edited by M. Thoma and A. Wyner

For information about Vols. 1-96 please contact your bookseller or Springer-Verlag

# Lecture Notes in Control and Information Sciences

Edited by M. Thoma and A. Wyner

## 166

## L. L. M. van der Wegen

# Local Disturbance Decoupling with Stability for Nonlinear Systems

Springer-Verlag
Berlin Heidelberg GmbH

**Author**
Dr. Leonardus Ludovicus Marie van der Wegen
School of Management Studies
University of Twente
P.O. Box 217
7500 AE Enschede
The Netherlands

ISBN 978-3-540-54543-9          ISBN 978-3-540-38461-8 (eBook)
DOI 10.1007/978-3-540-38461-8

## PREFACE

In the last decade quite some research has been carried out on stabilization problems for nonlinear control systems on the one side and to synthesis problems on the other. Until now, little attention has been paid to the study of systems for which a design objective as well as a stability requirement have to be met at the same time. This monograph fills up part of this gap by developing a local theory for the disturbance decoupling problem with stability.

I gratefully acknowledge the support of Henk Nijmeijer and Arjan van der Schaft who introduced me in the field of nonlinear systems theory and were always willing to lend me an ear. Furthermore, I would like to thank Henri Huijberts for many helpful discussions and for his contribution to the solution of the problems treated in Chapter 6 and Jessy Grizzle for his valuable comments on an earlier version of the manuscript.

Enschede, June 1991                                            Leo van der Wegen

# CONTENTS

# 1. INTRODUCTION

In this monograph the Disturbance Decoupling Problem with Stability for nonlinear systems is treated. In general terms this problem can be formulated as follows. Consider the system configuration in Figure 1.

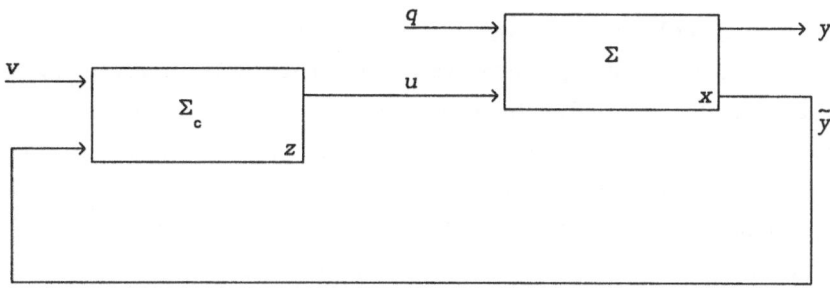

**Figure 1.** A feedback control system

Given is a system $\Sigma$ (which may be linear or nonlinear, finite- or infinite-dimensional) which is influenced by two kinds of inputs: the controlled inputs $u$ and the (uncontrolled) disturbance inputs $q$. The system has two kinds of outputs: the to-be-controlled outputs $y$ and the measurements $\tilde{y}$ of the state variables $x$. The Disturbance Decoupling Problem with Stability is defined as follows.

Find - if possible - a compensator $\Sigma_c$ with state variables $z$, outputs $u$, inputs $\tilde{y}$ and new controlled inputs $v$, such that for the feedback system the disturbances $q$ do not influence the outputs $y$ while the feedback system is stable. We do not yet formalize the stability requirement here, but one may think of asymptotic or exponential stability of the internal dynamics of the feedback system or bounded-disturbance-bounded-state stability where the disturbances $q$ are considered as inputs. The new controlled inputs $v$ are assumed to be present in the feedback system in order that other design specifications (e.g. reference signal tracking) may be met.

If the dimension of the compensator state vector is larger than zero, we speak about *dynamic* disturbance decoupling. If this dimension equals zero the compensator reduces to a *static* feedback $u = u(\tilde{y},v)$, and we speak about static disturbance decoupling, or simply disturbance decoupling if no confusion is possible.

Disturbance decoupling is a typical design problem in the sense that it plays a role in various other problems like model matching and reference signal tracking. For linear systems disturbance decoupling or at least minimizing the influence of the disturbances on the to-be-controlled outputs (in some suitable norm) is fundamental to $H_2$- or $H_\infty$-optimal control.

For linear systems the Disturbance Decoupling Problem with Stability has been completely solved, see e.g. the standard textbook [Wo]. In this book also other kinds of design problems are fruitfully treated by applying geometric methods. Motivated by this, at the end of the 70s researchers in nonlinear systems theory started to translate these methods to a nonlinear context using differential-geometric tools. This led to results on disturbance decoupling (see e.g. [Hi2], [IKGM1], [NvdS1], [NvdS2]), input-output decoupling (e.g. [SR], [Fr], [Si1], [Si2], [NS], [NR], [XG]), feedback linearization (e.g. [Br], [JR], [HSuM]) and invertibility (e.g. [Hi1], [Nij1], [Fl], [RN]).
By now, the *local* solution of the Disturbance Decoupling Problem (without stability) for nonlinear systems is well understood, at least if some constant rank assumptions hold (see e.g. [Is], [NvdS4]). Nevertheless, very few articles have been written about the solution of the (Local) Disturbance Decoupling Problem with Stability (see [BI2], [vdWN1], [vdWN2], [vdW1], [vdW2]), although the general problem of stabilizability of nonlinear control systems is classic (see e.g. [AG], [Gr]). This problem received revived attention in the last decade (see e.g. [Ae1], [Ae2], [Ba], [BI3], [So], [Ts], [VV]).

First, we consider in this Introduction the Disturbance Decoupling Problem and the Disturbance Decoupling Problem with Stability for linear systems to make the reader familiar with the geometric approach in general and with some specific subspaces in particular that play a key role in the solution of these problems. As a matter of fact, most of the theory in Chapters 3 and 4 is concerned with finding a proper generalization of these subspaces for nonlinear systems.
Consider the linear system

(1.1) $\quad \begin{cases} \dot{x} = Ax + Bu + Eq, \ x \in \mathbb{R}^n, \ u \in \mathbb{R}^m, \ q \in \mathbb{R}^r \\ y = Cx, \qquad\qquad y \in \mathbb{R}^\ell \end{cases}$

where A, B, E and C are matrices of appropriate dimensions. The absence of an equation for $\tilde{y}$ in (1.1) means that we assume that all states $x$ are measurable. Hence, we may apply state feedback rather than output feedback. This will be a standing assumption throughout this monograph. Moreover, for the time being, we assume that we apply static feedback.

**Definition 1.1** A feedback

$$(1.2) \qquad u = Fx + Gv, \qquad u \in \mathbb{R}^m, \ v \in \mathbb{R}^m, \ x \in \mathbb{R}^n$$

with $F$ and $G$ matrices of appropriate size is called a *regular static state feedback* if $G$ is a nonsingular matrix.

This implies that the feedback system (1.1,2) admits as many independent controls ($m$) as the original system (1.1). The Disturbance Decoupling Problem and the Disturbance Decoupling Problem with Stability are defined as follows.

**Definition 1.2** Consider the system (1.1).
(i) *Disturbance Decoupling Problem* (DDP)
Under what conditions can we find a regular static state feedback (1.2) such that in the feedback system (1.1,2) the disturbances $q$ do not influence the outputs $y$?
(ii) *Disturbance Decoupling Problem with Stability* (DDPS)
Under what conditions can we find a regular static state feedback (1.2) such that for the feedback system (1.1,2) the DDP is solved and the modified drift dynamics $\dot{x} = (A+BF)x$ are asymptotically stable?

Note that the modified drift dynamics are exactly the dynamics of (1.1,2) with the new inputs $v$ and the disturbances $q$ equal to zero.
Suppose that the DDP is solvable for (1.1,2). Then for $x(0) = 0$ and $v \equiv 0$ the outputs $y$ given by

$$(1.3) \qquad y(t) = C \int_0^t e^{(A+BF)(t-s)} Eq(s)ds$$

are identically equal to zero for all $t$. This implies that the subspace

$$(1.4) \qquad \mathcal{V} := \text{im } E + (A+BF)\text{im } E + \ldots + (A+BF)^{n-1}\text{im } E =: \ < A+BF \mid \text{im } E >$$

is contained in ker $C$. Moreover, $V$ is $(A+BF)$-invariant, i.e. $(A+BF)V \subset V$. Such a subspace $V$ is called controlled invariant.

**Definition 1.3(a)** ([BM1], [Wo], [Wi]) Consider the system $\dot{x} = Ax + Bu$ with $x \in \mathbb{R}^n$, $u \in \mathbb{R}^m$. A subspace $V$ in $\mathbb{R}^n$ is called *controlled invariant* (or $(A,B)$-invariant) if one of the following equivalent conditions holds:

(i)    For every $x_0 \in V$ there exists an admissible control $u(t)$ such that $x_u(t,x_0) \in V$ for all $t \geq 0$.

(ii)   There exists a feedback $u = Fx$ such that $(A+BF)V \subset V$.

(iii)  $AV \subset V + \text{im } B$.

A matrix $F$ that fulfills (ii) is called a "friend" of $V$, denoted as $F \in \mathcal{F}(V)$.

In (i) $x_u(t,x_0)$ denotes the state trajectory of the system $x = Ax + Bu$ with $x(0) = x_0$. A control $u(t)$ is said to be admissible if the trajectory $x_u(t,x_0)$ is absolutely continuous.

For later use we define the dual notion of $(A,B)$-invariance here.

**Definition 1.3(b)** ([BM1], [Sch]) Consider the system $\dot{x} = Ax$, $y = Cx$ with $x \in \mathbb{R}^n$, $y \in \mathbb{R}^\ell$. A subspace $V$ in $\mathbb{R}^n$ is called *conditioned invariant* (or $(C,A)$-invariant) if one of the following equivalent conditions holds:

(i)    There exists a mapping $G$ such that $(A+GC)V \subset V$.

(ii)   $A(V \cap \text{ker } C) \subset V$.

Returning to the DDP it follows that its solvability implies that there exists a controlled invariant subspace $V$ such that

(1.5)      $\text{im } E \subset V \subset \text{ker } C$

As a matter of fact, the following result holds:

**Theorem 1.4** ([Wo]) Consider the system (1.1). The DDP for (1.1) is solvable if and only if

(1.6)      $\text{im } E \subset V^*$

where $V^*$ denotes the largest controlled invariant subspace in ker $C$ (which exists!).

Actually, $V^*$ can be calculated explicitly using the following algorithm.

**Algorithm 1.5** ([Wo])

(1.7)     $V_0 := \ker C, \quad V_{k+1} := \ker C \cap A^{-1}(V_k + \operatorname{im} B), \quad V^* := V_n$

Loosely speaking, the DDP for (1.1) is solvable if and only if there exists a subspace $V$ in the kernel of $C$ in which the disturbances are caught and that can be made invariant by feedback (so the disturbances stay captured and cannot influence the outputs). Note that the second part of the feedback ($Gv$) is not important for the solution of the DDP and is added in order that other design specifications may be met simultaneously.

We now turn to the solution of the DDPS. Obviously, a necessary condition for solvability of the DDPS is that the system (1.1) is stabilizable. Suppose now that the DDPS for (1.1) is solvable. Then there exists a feedback (1.2) that makes some controlled invariant subspace $V$ with $\operatorname{im} E \subset V \subset \ker C$ invariant and, moreover, it makes that the drift dynamics of the feedback system (1.1,2) are asymptotically stable and so also the drift dynamics restricted to $V$ are asymptotically stable, i.e. if $x_0 \in V$, then the solution $x(t)$ of the differential equations $\dot{x} = (A+BF)x$, $x(0) = x_0$ is such that $x(t) \in V$ for all $t \geq 0$ and $x(t) \to 0$ if $t \to \infty$. In fact, $V$ is a stabilizability subspace.

**Definition 1.6** ([Hau], [Tr]) Consider the system $\dot{x} = Ax + Bu$, with $x \in \mathbb{R}^n$, $u \in \mathbb{R}^m$. A subspace $V$ in $\mathbb{R}^n$ is called a *stabilizability* subspace if one of the following equivalent conditions holds:
(i)  For every $x_0 \in V$ there exists an admissible control $u(t)$ such that $x_u(t,x_0)$ is Bohl, $x_u(t,x_0) \in V$ for all $t \geq 0$ and $x_u(t,x_0) \to 0$ if $t \to \infty$.
(ii) There exists an $F \in \mathcal{F}(V)$ such that $\sigma\big((A+BF)|_V\big) \subset \mathbb{C}^-$, the open left-half of the complex plane.

As usual, $\sigma(K)$ denotes the set of eigenvalues of a matrix $K$. Note that, in particular, a stabilizability subspace is controlled invariant. It is well-known (see e.g. [Wo]) that by making a subspace invariant, one partly fixes the drift dynamics of the system restricted to that subspace. This diminishes the possibilities of achieving stability and disturbance decoupling at the same time. Stabilizability subspaces are exactly those controlled invariant subspaces $V$ for which these drift dynamics (1.1,2) restricted to $V$ can be stabilized asymptotically with some $F \in \mathcal{F}(V)$. The solution of the DDPS is given by the following theorem.

**Theorem 1.7** ([Wo]) The DDPS for the stabilizable system (1.1) is solvable if and only if

(1.8)     $\text{im } E \subset V_s^*$

where $V_s^*$ denotes the largest stabilizability subspace in $\ker C$ (which exists!).

Before pointing out how $V_s^*$ can be calculated, we define the concept of controllability subspace.

**Definition 1.8** ([Wo], [Wi]) Consider the system $\dot{x} = Ax + Bu$ with $x \in \mathbb{R}^n$, $u \in \mathbb{R}^m$. A subspace $V$ in $\mathbb{R}^n$ is called a *controllability* subspace if one of the following equivalent conditions holds:
(i)     For every $x_0$ and $x_1 \in V$ there exist a $T > 0$ and an admissible control $u(t)$ such that $x_u(t,x_0) \in V$ for all $t \geq 0$ and $x_u(T,x_0) = x_1$.
(ii)    There exist linear maps $F: \mathbb{R}^n \to \mathbb{R}^m$ and $G: \mathbb{R}^m \to \mathbb{R}^m$ such that $V = < A+BF| \text{ im } BG >$.
(iii) For any $F \in \mathcal{F}(V)$ we have $V = < A+BF| \text{ im } B \cap V >$.

**Lemma 1.9** ([Wo]) Consider the system (1.1). Then the largest controllability subspace in $\ker C$, denoted by $\mathcal{R}^*$, exists. Moreover, $\mathcal{R}^* \subset V^*$ and if $F \in \mathcal{F}(V^*)$, then $F \in \mathcal{F}(\mathcal{R}^*)$.

Starting from $V^*$ the subspace $\mathcal{R}^*$ can be computed as follows.

**Algorithm 1.10** ([Wo])

(1.9)     $\mathcal{R}_0 := \text{im } B \cap V^*$,     $\mathcal{R}_{k+1} := (A\mathcal{R}_k + \text{im } B) \cap V^*$,     $\mathcal{R}^* := \mathcal{R}_n$

Now the subspace $V_s^*$ can be found in the following way. Choose a feedback (1.2) for system (1.1) with $F \in \mathcal{F}(V^*)$. Since the dynamics of the system $\dot{x} = (A+BF)x + BGv$ restricted to $\mathcal{R}^*$ are controllable, the feedback (1.2) can be chosen in such a way that $\sigma\big((A+BF)|_{\mathcal{R}^*}\big) \subset \mathbb{C}^-$. The eigenvalues of the matrix $(A+BF)|_{V^*/\mathcal{R}^*}$ are fixed (independent of $F \in \mathcal{F}(V^*)$). These eigenvalues correspond to the well-known transmission zeros (see [Wo]) if the system

(1.1) is controllable and observable. Let $\mathcal{X}^- \subset \mathcal{V}^*/\mathcal{R}^*$ denote the eigenspace corresponding to the exponentially stable transmission zeros. Then $\mathcal{V}_s^*$, the largest stabilizability subspace in ker $C$, is equal to $P^{-1}\mathcal{X}^- \subset \mathcal{V}^*$, where $P$ denotes the canonical projection $P: \mathbb{R}^n \to \mathbb{R}^n/\mathcal{R}^*$.

**Remark 1.11** In the definition of the DDPS it is required that the feedback system (1.1,2) has asymptotically (or, equivalently, exponentially) stable drift dynamics. This automatically implies that the system (1.1,2) with $x(0) = 0$ and $v = 0$ fulfills the condition $\|x(\cdot)\| \le \gamma\|q(\cdot)\|$ for some constant $\gamma$ ($\|(\cdot)\|$ denotes some norm on the space of disturbance/state functions). Hence, the system (1.1,2) is bounded–disturbance bounded–state stable (BDBS–stable). □

In the sequel we consider nonlinear control systems

$$
(1.10) \quad
\begin{cases}
\dot{x} = f(x) + g(x)u + p(x)q := f(x) + \sum_{i=1}^{m} g_i(x)u_i + \sum_{i=1}^{r} p_i(x)q_i, \ x \in \mathbb{R}^n \\
y = h(x) = \bigl(h_1(x), \dots, h_\ell(x)\bigr)^T
\end{cases}
$$

where $x$, $u$, $q$ and $y$ denote the states, inputs, disturbances and outputs, respectively, and $g(x)$ and $p(x)$ are matrices of dimension $n \times m$ and $n \times r$ with columns $g_i(x)$, $i = 1, \dots, m$ and $p_j(x)$, $j = 1, \dots, r$, respectively. The vector fields $f$ and $g_i$, $i = 1, \dots, m$ and $p_j$, $j = 1, \dots, r$, and the output functions $h_i$, $i = 1, \dots, \ell$, are assumed to be smooth, i.e. infinitely many times continuously differentiable (see Chapter 2). Note that the system (1.10) is affine in the inputs $u$ and the disturbances $q$.

**Definition 1.12** A feedback

$$(1.11) \quad u = \alpha(x) + \beta(x)v, \quad u \in \mathbb{R}^m, \ v \in \mathbb{R}^m, \ x \in \mathbb{R}^n$$

with $\alpha: \mathbb{R}^n \to \mathbb{R}^m$ and $\beta: \mathbb{R}^n \to \mathbb{R}^{m \times m}$ is called a *regular static state feedback* if $\beta(x)$ is a nonsingular matrix for all $x$.

The feedback system (1.10,11) has the form

$$
(1.12) \quad
\begin{cases}
\dot{x} = f(x) + g(x)\alpha(x) + g(x)\beta(x)v + p(x)q \\
y = h(x)
\end{cases}
$$

The Disturbance Decoupling Problem is defined as follows (cf. [Hi2], [IKGM1], [NvdS2]).

**Definition 1.13** *Disturbance Decoupling Problem* (DDP)
Consider the smooth nonlinear system (1.10). Under what conditions can we find a smooth regular static state feedback (1.11) such that in the feedback system (1.10,11) the disturbances $q$ do not influence the outputs $y$?

Note that the decoupling requirement must hold for all initial points $x_0$ and all controlled inputs $v$! The DDP is of a global nature, because the system (1.10) and the feedback (1.11) are globally defined. Easily verifiable conditions for the solvability of this global problem are not known at the moment. Therefore, we consider only the Local Disturbance Decoupling Problem, abbreviated by LDDP. Here "local" refers to the fact that we search for a feedback defined on a neighborhood $\mathcal{O}$ of a given point such that the disturbance decoupling requirement holds for all initial points in $\mathcal{O}$ and all controlled inputs $v$, as long as the state trajectories remain within $\mathcal{O}$.

**Definition 1.14** *Local Disturbance Decoupling Problem* (LDDP) ([Hi2], [Is], [NvdS4]) Consider the smooth system (1.10) and a point $x_0 \in \mathbb{R}^n$. Under what conditions can we find a smooth regular static state feedback (1.11) defined locally around $x_0$ such that in the feedback system (1.10,11) the disturbances $q$ do not influence the outputs $y$?

**Remark 1.15** For linear systems solvability of the DDP(S) does not depend on the choice of the matrix $G$ in the feedback (1.2). As a matter of fact, if the regular feedback (1.2) solves the DDP(S) for (1.1), then disturbance decoupling (with stability) can also be obtained by applying the feedback $u = Fx + \tilde{G}v$ for any arbitrary matrix $\tilde{G}$ (possibly singular).
However, for nonlinear systems invertibility of $\beta(x)$ is *not* a nontrivial restriction as follows from the following example:

$$(1.13) \quad \dot{x}_1 = -x_1 + q, \ \dot{x}_2 = -x_2 + x_1 u, \ y = x_2$$

Obviously, the control $u \equiv 0$ decouples the output in (1.13) from the disturbance. Hence, disturbance decoupling in (1.13) can be obtained by applying a singular feedback. On the other hand, it seems impossible to find a smooth regular state feedback (1.11) that solves the LDDP for this system locally around $x = 0$. This follows from the fact that such a regular feedback should contain a term of the form $1/x_1$. Clearly, similar arguments apply to the disturbance decoupling problem including a stability requirement. ($u = 0$ solves a singular version of the LDDPS as formulated in Definition 1.18.)                                                                     □


The solution of the LDDP is given in Chapter 2. To define the (Local) Disturbance Decoupling Problem with Stability it is necessary to decide what kind of stability is desirable for the feedback system (1.12). Roughly speaking, there are three possibilities (assume that $f(x_0) = 0$):

(i)   the modified drift dynamics $\dot{x} = (f+g\alpha)(x)$ are locally asymptotically stable around $x_0$;

(ii)  the modified drift dynamics are locally exponentially stable around $x_0$, i.e. the linearized system $\dot{z} = \left(\frac{\partial f}{\partial x}(x_0) + g(x_0)\frac{\partial \alpha}{\partial x}(x_0)\right)z$ is asymptotically stable;

(iii) the system (1.12) is locally BDBS-stable, i.e. there exist neighborhoods $\mathcal{O}$ and $\tilde{\mathcal{O}}$ of $x_0$ and a constant $D$ such that if $v = 0$, $x_0 \in \tilde{\mathcal{O}}$ and $q(t)$ pointwise bounded by $D$, then $x(t) \in \mathcal{O}$ for all $t \geq 0$.


Note that local BDBS-stability as defined here is a local version of more general stability concepts such as input-to-state stability ([So]) and total stability (see e.g. [Ha]).

Since we are looking for a solution of the disturbance decoupling problem locally around a given point $x_0 \in \mathbb{R}^n$, it is natural to require that the feedback system is locally asymptotically or exponentially stable, because this implies that the state trajectories remain bounded. Furthermore, local exponential stability implies local BDBS-stability (see Lemma 1.16). On the other hand, a BDBS-stable system that is not asymptotically stable may give rise to undesired behavior, as is illustrated by Example 1.17. For these reasons we will require the feedback system to be exponentially stable around the equilibrium $x_0$. In the sequel we take $x_0 = 0$ for convenience.


**Lemma 1.16** Consider the smooth nonlinear system (1.10). Assume that $x = 0$ is a locally exponentially stable equilibrium of $f$. Suppose that the vector fields $p_i$, $i = 1,\ldots,r$ are bounded. Then the system (1.10) is locally BDBS-stable. Precisely, consider the system (1.10) with $u = 0$ and the disturbances as inputs. Then there exist neighborhoods $O$ and $\tilde{O}$ of $x = 0$ and a constant $D$ such that if $x_0 \in \tilde{O}$ and $|q(t)| \leq D$ for all $t$ ($|\ |$ denotes the Euclidean norm), then $x(t) \in O$ for all positive $t$.

This result is well-known (cf. Theorem 1 in [So] and Section 56 in [Ha]). The proofs in [So] and [Ha] make use of inverse Lyapunov theorems. A straightforward proof of Lemma 1.16 using Gronwall's lemma is given in Appendix A.

**Example 1.17** ([HSM]) Consider the following model of an aircraft

$$(1.14) \quad \begin{cases} \dot{x}_1 = x_2 \\ \dot{x}_2 = -\sin(\theta_1)u_1 + \epsilon\cos(\theta_1)u_2 + p_1(x,y)q \\ \dot{y}_1 = y_2 \\ \dot{y}_2 = -1 + \cos(\theta_1)u_1 + \epsilon\sin(\theta_1)u_2 + p_2(x,y)q \quad , \quad \epsilon > 0 \\ \dot{\theta}_1 = \theta_2 \\ \dot{\theta}_2 = u_2 \end{cases}$$

with outputs

$$(1.15) \quad z_1 = x_1, \quad z_2 = y_1$$

Choosing in (1.14) the feedback

$$(1.16) \quad \begin{bmatrix} u_1 \\ u_2 \end{bmatrix} = \begin{bmatrix} -\sin(\theta_1) & \cos(\theta_1) \\ \epsilon^{-1}\cos(\theta_1) & \epsilon^{-1}\sin(\theta_1) \end{bmatrix} \left\{ \begin{bmatrix} 0 \\ 1 \end{bmatrix} + \begin{bmatrix} -x_1-2x_2 \\ -y_1-2y_2 \end{bmatrix} + \begin{bmatrix} v_1 \\ v_2 \end{bmatrix} \right\}$$

yields

$$(1.17a) \quad \begin{cases} \dot{x}_1 = x_2 \\ \dot{x}_2 = -x_1-2x_2 + v_1 + p_1(x,y)q \\ \dot{y}_1 = y_2 \\ \dot{y}_2 = -y_1-2y_2 + v_2 + p_2(x,y)q \end{cases}$$

and

$$(1.17b) \quad \begin{cases} \dot{\theta}_1 = \theta_2 \\ \dot{\theta}_2 = \epsilon^{-1}\sin(\theta_1) + \epsilon^{-1}\cos(\theta_1)(-x_1-2x_2+v_1) + \epsilon^{-1}\sin(\theta_1)(-y_1-2y_2+v_2) \end{cases}$$

The input-output behavior of the system is fully described by the equations (1.17a) and (1.15). It immediately follows from Lemma 1.16 that in case

$v = 0$ and $p_1$ and $p_2$ are bounded functions of $x_1$, $x_2$, $y_1$, $y_2$, then the system (1.17a,15) is locally BDBS-stable if the disturbances $q$ are considered as inputs. Nevertheless, the behavior of the overall system (1.17,15) is not satisfactory. This can be seen as follows. If $v = 0$ and $q = 0$ then $x_1$, $x_2$, $y_1$ and $y_2$ tend to zero if $t$ tends to infinity. Hence, the asymptotic behavior of the overall system is determined by

$$(1.18) \qquad \begin{cases} \dot{\theta}_1 = \theta_2 \\ \dot{\theta}_2 = \epsilon^{-1}\sin(\theta_1) \end{cases}$$

These dynamics are not asymptotically stable. As a matter of fact, around the equilibrium $(\theta_1, \theta_2) = (0,0)$ the system is exponentially unstable. Equation (1.18) implies that the aircraft will (depending on the initial conditions) either rock from side to side or roll continuously in one direction (except at the equilibrium point).  □

We end this chapter with the formulation of the Local Disturbance Decoupling Problem with Stability, abbreviated by LDDPS.

**Definition 1.18** *Local Disturbance Decoupling Problem with Stability* (LDDPS) Consider the smooth nonlinear system (1.10) with $f(0) = 0$. Under what conditions can we find a smooth regular static state feedback (1.11) defined locally around $x = 0$ with $\alpha(0) = 0$ such that in the feedback system (1.10,11) the disturbances $q$ do not influence the outputs $y$, and $x = 0$ is a locally exponentially stable equilibrium of the modified drift dynamics $\dot{x} = f(x)+g(x)\alpha(x)$?

**Remark 1.19**
(i) We use the abbreviations DDP and DDPS to refer throughout to the versions of the disturbance decoupling problems for linear systems and LDDP and LDDPS to the *local* versions of these problems for nonlinear systems.
(ii) It follows from the definition of the LDDPS that this problem is solvable only if the pair $\left(\frac{\partial f}{\partial x}(0), g(0)\right)$ is stabilizable.  □

**Organization**
The rest of this monograph is organized as follows. In Chapter 2 some basic definitions from differential geometry are recalled (Section 2.2) and the solution of the Local Disturbance Decoupling Problem is given (Section

2.3). For readers who are familiar with this theory Section 2.4 gives a quick overview of the basic definitions and abbreviations used in the sequel and defined earlier in Chapter 2. Section 2.5 contains some material on accessibility and on constrained and zero dynamics. In this section a new notion of restricted zero dynamics is defined. Finally, some notions from dynamical systems are summarized.

In Chapters 3 and 4 two methods are considered to solve the Local Disturbance Decoupling Problem with Stability. In Section 4.3 a closer look is taken at the conditions under which the LDDPS has been solved. In Section 4.4 the results obtained earlier are compared to the results given in [BI2].

The solvability of the LDDPS for a nonlinear system in connection with the solvability of the DDPS for its linearization is studied in Chapter 5. Attention is paid to the relation between the feedbacks that solve the DDPS and the LDDPS, respectively, and to solvability of the LDDPS for a nonlinear system by applying a linear feedback.

In Section 6.1 it is shown by means of an example that the class of systems for which the outputs can be decoupled from the disturbances becomes larger if one is allowed to apply dynamic feedback. This gives rise to the definition of the Local Dynamic Disturbance Decoupling Problem and the Local Dynamic Disturbance Decoupling Problem with Stability, treated in Sections 6.3 and 6.4, respectively. In Section 6.2 Singh's algorithm, which plays a key role in the solution of these problems, is recalled.

Finally, in Chapter 7 conclusions are drawn and some open problems are mentioned.

## 2. PRELIMINARIES

### 2.1  Introduction

In this chapter we present background material. More detailed information
can be found in the literature cited in each section as well as in
references therein. The organization of this chapter is as follows. Section
2.2 gives an overview of the concepts from differential geometry used in
this monograph. In Section 2.3 the notion of (local) controlled invariance
is introduced and the solution of the LDDP is given. Moreover, related
concepts like relative degrees, decoupling matrix and controllability
distributions are defined. In the last section the notions of (strong)
accessibility and constrained and zero dynamics are recalled and the new
notion of restricted zero dynamics is introduced. Furthermore, some results
from the theory of dynamical systems are given, especially on the existence
of certain invariant manifolds. In Section 2.4 the problem formulation of
the LDDPS and the main results from the previous sections are summarized.

### 2.2  Basic definitions from differential geometry

Standard references on differential geometry are [Bo], [Sp]. The introduction of concepts from differential geometry given in this section closely
follows [Ak]. Some notions are taken from [Is], [NvdS4] and [Sp].
Consider the space $\mathbb{R}^n$. The *tangent space* $T_x\mathbb{R}^n$ of $\mathbb{R}^n$ in $x$ is the set of
vectors that are tangent to $\mathbb{R}^n$ in $x \in \mathbb{R}^n$ (and so, $T_x\mathbb{R}^n$ is a copy of $\mathbb{R}^n$).
The elements of $T_x\mathbb{R}^n$ are called *tangent vectors*. The natural basis of $T_x\mathbb{R}^n$
will be denoted by $\{\frac{\partial}{\partial x_1}\big|_x, \ldots, \frac{\partial}{\partial x_n}\big|_x\}$.
A *vector field* $f$ on $\mathbb{R}^n$ is a mapping assigning to each point $x \in \mathbb{R}^n$ a tan-
gent vector $f(x) = (f_1(x), \ldots, f_n(x))^T = \sum_{i=1}^{n} f_i(x)\frac{\partial}{\partial x_i}\big|_x$ (where $T$ denotes
transpose) in $T_x\mathbb{R}^n$. $f$ is a *smooth* ($C^\infty$) vector field if the component
functions $f_1, \ldots, f_n$ are smooth functions (i.e. functions that are
infinitely many times continuously differentiable). $f$ is *analytic* if the
component functions are analytic. $V(\mathbb{R}^n)$ denotes the set of all smooth
vector fields on $\mathbb{R}^n$. This set is a vector space over $\mathbb{R}$ and, moreover, a

*Lie-algebra* with the *Lie-bracket* defined as follows: If $f$ and $g$ are two smooth vector fields, then $[f,g]$ is a smooth vector field given by

(2.2.1)    $[f,g] := \dfrac{\partial g}{\partial x} f - \dfrac{\partial f}{\partial x} g$

**Remark 2.2.1**  $(V,[\ ,\ ])$ is a Lie-algebra if $V$ is a vector space and the binary operation $[\ ,\ ]: V \times V \to V$ has the following properties:

(i)    it is skew-symmetric, i.e. $[v,w] = -[w,v]$;

(ii)   it is bilinear over $\mathbb{R}$, i.e. $[a_1 v_1 + a_2 v_2, w] = a_1[v_1,w] + a_2[v_2,w]$, $a_1, a_2 \in \mathbb{R}$;

(iii)  it satisfies the Jacobi-identity

(2.2.2)    $\Big[v,[w,z]\Big] + \Big[w,[z,v]\Big] + \Big[z,[v,w]\Big] = 0$                              □

Instead of $[f,g]$ the notation $\mathrm{ad}_f g$ is used. The latter may be used iteratively. Hence,

(2.2.3)    $\mathrm{ad}_f^0 g = g$,   $\mathrm{ad}_f g = [f,g]$,   $\mathrm{ad}_f^{k+1} g = \mathrm{ad}_f(\mathrm{ad}_f^k g)$,   $k = 1,2,\ldots$

The dual space of $T_x \mathbb{R}^n$, the *cotangent space* of $\mathbb{R}^n$ in $x$ is denoted by $T_x^* \mathbb{R}^n$. Its elements, called *tangent covectors*, are by definition linear functionals on $T_x \mathbb{R}^n$. If $v^* \in T_x^* \mathbb{R}^n$, then the value of $v^*$ at $v \in T_x \mathbb{R}^n$ is denoted by $<v^*, v>$. The dual basis of $\{\frac{\partial}{\partial x_1}\big|_x, \ldots, \frac{\partial}{\partial x_n}\big|_x\}$ is denoted by the tangent covectors $\{dx_1\big|_x, \ldots, dx_n\big|_x\}$.

A *covector field* (or one-form) on $\mathbb{R}^n$ is a mapping assigning to each point $x \in \mathbb{R}^n$ a tangent covector $\omega(x) = (\omega_1(x), \ldots, \omega_n(x)) = \sum_{i=1}^{n} \omega_i(x) dx_i\big|_x$ in $T_x^* \mathbb{R}^n$. $\omega$ is smooth if the functions $\omega_1, \ldots, \omega_n$ are.

With every $\lambda \in C^\infty(\mathbb{R}^n)$, the set of smooth functions on $\mathbb{R}^n$, we can associate a tangent covector $d\lambda$ defined by $d\lambda(x) = \sum_{i=1}^{n} \frac{\partial \lambda}{\partial x_i}(x) dx_i\big|_x$. If $\omega$ is a covector field and $f$ a vector field, then the dual product of $\omega$ and $f$, written as $<\omega, f>$, denotes the function defined by $<\omega, f>(x) = <\omega(x), f(x)>$. The set of all smooth covector fields on $\mathbb{R}^n$ is denoted by $V^*(\mathbb{R}^n)$.

**Definition 2.2.2**  Let $f \in V(\mathbb{R}^n)$. The following *Lie-derivatives* may be related to $f$:

(i)    $\lambda \in C^\infty(\mathbb{R}^n)$:   $L_f : C^\infty(\mathbb{R}^n) \to C^\infty(\mathbb{R}^n)$

$$(2.2.4) \quad L_f \lambda = <d\lambda, f> = \sum_{i=1}^{n} \frac{\partial \lambda}{\partial x_i} f_i$$

(ii) $g \in V(\mathbb{R}^n)$: $\quad$ $ad_f : V(\mathbb{R}^n) \to V(\mathbb{R}^n)$

$$(2.2.5) \quad ad_f g = [f,g] \quad (cf. \ (2.2.1))$$

(iii) $\omega \in V^*(\mathbb{R}^n)$: $\quad$ $L_f : V^*(\mathbb{R}^n) \to V^*(\mathbb{R}^n)$

$$(2.2.6) \quad L_f \omega = \left[ \frac{\partial \omega^T}{\partial x} f \right]^T + \omega \frac{\partial f}{\partial x}$$

The three types of Lie-derivatives are related by the *Leibnitz formula*

$$(2.2.7) \quad L_f <\omega, g> = <L_f \omega, g> + <\omega, ad_f g>$$

A *distribution* $\Delta$ on $\mathbb{R}^n$ is a rule assigning to each $x \in \mathbb{R}^n$ a subspace $\Delta(x) \subset T_x \mathbb{R}^n$ such that for every $x \in \mathbb{R}^n$, there exist a neighborhood $\mathcal{O}(x)$ of $x$ and a set of vector fields defined on $\mathcal{O}(x)$ denoted $\{f_i | i \in I\}$ with the property that $\Delta(y) = sp\{f_i(y) | i \in I\}$ for all $y \in \mathcal{O}(x)$. The distribution is smooth if one can choose smooth vector fields $\{f_i | i \in I\}$.
If $\{g_j | j \in J\}$ is a set of smooth vector fields defined on $\mathbb{R}^n$, then their span, denoted by $sp\{g_j | j \in J\}$ is the smooth distribution defined by

$$(2.2.8) \quad sp\{g_j | j \in J\}: x \mapsto sp\{g_j(x) | j \in J\}$$

The sum and intersection of two distributions $\Delta_1$ and $\Delta_2$ are defined as

$$(2.2.9) \quad \Delta_1 + \Delta_2 : x \mapsto \Delta_1(x) + \Delta_2(x)$$

$$(2.2.10) \quad \Delta_1 \cap \Delta_2 : x \mapsto \Delta_1(x) \cap \Delta_2(x)$$

Note that the sum of two smooth distributions is smooth again, but the intersection need not be smooth. However, if $\Delta$ is a distribution there always exists a largest smooth distribution $smt(\Delta)$ contained in $\Delta$. As a matter of fact, $smt(\Delta)$ is exactly equal to $sp\{X | X$ smooth vector field in $\Delta\}$. A vector field $f$ belongs to a distribution $\Delta$, denoted as $f \in \Delta$, if $f(x) \in \Delta(x)$ for all $x \in \mathbb{R}^n$. If $\Delta_1$ and $\Delta_2$ are two distributions, then $\Delta_1$ is contained in $\Delta_2$, denoted as $\Delta_1 \subset \Delta_2$, if any vector field in $\Delta_1$ belongs to $\Delta_2$. A distribution $\Delta$ on $\mathbb{R}^n$ is said to be *nonsingular* if $dim \ \Delta(x) = d$ for all $x \in \mathbb{R}^n$. The constant $d$ is called the dimension of $\Delta$. A set of vector fields $\{g_i | i = 1, \ldots, r\}$ is called *independent* if $sp\{g_1(x), \ldots, g_r(x)\} = r$ for all $x$.

If $\Delta$ is a smooth nonsingular $d$-dimensional distribution on $\mathbb{R}^n$, then for any $x \in \mathbb{R}^n$ there exist a neighborhood $\mathcal{O}(x)$ of $x$ and a set $\{f_1, \ldots, f_d\}$ of smooth vector fields defined on $\mathcal{O}(x)$ with the property that for all $y \in \mathcal{O}(x)$ $\Delta(y) = \mathrm{sp}\{f_1(y), \ldots, f_d(y)\}$.

A distribution $\Delta$ is called *involutive* if the Lie-bracket $[f_1, f_2]$ of any pair of vector fields $f_1$ and $f_2$ belonging to $\Delta$ is a vector field which again belongs to $\Delta$.

In case $\Delta$ is not involutive, there always exists a smallest involutive distribution containing $\Delta$. This distribution, called the *involutive closure* of $\Delta$ and denoted as inv clos($\Delta$), is in fact the intersection of all involutive distributions containing $\Delta$.

A mapping $z = \varphi(x)$ from an open set $\mathcal{O}$ in $\mathbb{R}^n$ to $\mathbb{R}^n$ is called a *coordinate transformation* if $\varphi$ is a diffeomorphism, i.e. $\varphi^{-1}$ exists and $\varphi$ as well as $\varphi^{-1}$ are smooth.

A nonsingular smooth $k$-dimensional distribution $\Delta$ on $\mathbb{R}^n$ is said to be *completely integrable* if at each $x \in \mathbb{R}^n$ there exist a neighborhood $\mathcal{O}(x)$ and a coordinate transformation $z = \varphi(x)$ defined on $\mathcal{O}(x)$ such that

$$(2.2.11) \quad \Delta(y) = \mathrm{sp}\{\frac{\partial}{\partial z_1}\Big|_y, \ldots, \frac{\partial}{\partial z_k}\Big|_y\}$$

for all $y \in \mathcal{O}(x)$.

A distribution $\Delta$ defined by (2.2.11) is said to be a *flat distribution* (in the coordinates $z_1, \ldots, z_n$).

**Theorem 2.2.3** *Frobenius' Theorem* (local version, see e.g. [NvdS4])
A nonsingular smooth distribution is completely integrable if and only if it is involutive.

A submanifold $M$ of $\mathbb{R}^n$ is an *integral manifold* of a distribution $\Delta$ on $\mathbb{R}^n$ if $T_x M = \Delta(x)$ for all $x \in M$. Now Frobenius' Theorem states that for smooth nonsingular distributions $\Delta$ involutivity is equivalent to the existence at each $x \in \mathbb{R}^n$ of a locally defined integral manifold of $\Delta$. Note that in case $\Delta$ is given by (2.2.11) the integral manifolds are given by

$$(2.2.12) \quad \{y \in \mathcal{O}(x) \mid y_{k+1} = c_{k+1}, \ldots, y_n = c_n, \ c_{k+1}, \ldots, c_n \in \mathbb{R}$$
$$\text{such that} \ (0, \ldots, 0, y_{k+1}, \ldots, y_n)^T \in \mathcal{O}(x)\}$$

A submanifold $M$ of $\mathbb{R}^n$ is called a *maximal integral manifold* if $M$ is connected and every other connected integral manifold containing $M$ coincides with $M$. A distribution $\Delta$ on $\mathbb{R}^n$ has the *maximal integral manifold property* if through every point $x \in \mathbb{R}^n$ passes a maximal integral manifold $S$ of $\Delta$ or, in other words, if there exists a partition of $\mathbb{R}^n$ into maximal integral manifolds of $\Delta$. Now, the global version of Theorem 2.2.3 reads as follows.

**Theorem 2.2.4** *Frobenius' Theorem* (global version, see e.g. [Sp])
A nonsingular smooth distribution has the maximal integral manifold property if and only if it is involutive.

The collection of maximal integral manifolds is called a *foliation* and any particular maximal integral manifold in this set a *leaf* of the foliation. Note that the collection of submanifolds (2.2.12) parametrized by $c_{k+1}, \ldots, c_n$ gives a locally defined foliation on $\mathcal{O}(x)$.

Another generalization of Theorem 2.2.3 deals with integrability of a set of nested distributions. A set of distributions $\{\Delta_1, \ldots, \Delta_r\}$ is called *nested* if $\Delta_1 \subset \Delta_2 \subset \ldots \subset \Delta_r$. A collection $\{\Delta_1, \ldots, \Delta_r\}$ of nested nonsingular smooth distributions on $\mathbb{R}^n$ is completely integrable if at each $x \in \mathbb{R}^n$ there exist a neighborhood $\mathcal{O}(x)$ and a coordinate transformation $z = \varphi(x)$ defined on $\mathcal{O}(x)$ such that for $i = 1, \ldots, r$, $\Delta_i(y) = \text{sp}\{\frac{\partial}{\partial z_1}\big|_y, \ldots, \frac{\partial}{\partial z_{d_i}}\big|_y\}$ for all $y \in \mathcal{O}(x)$ $(d_i = \dim(\Delta_i))$.

**Theorem 2.2.5** ([JR]) A collection $\{\Delta_1, \ldots, \Delta_r\}$ of nested nonsingular smooth distributions is completely integrable if and only if each distribution $\Delta_i$, $i = 1, \ldots, r$ is involutive.

A distribution $\Delta$ is said to be *regular* if $\Delta$ is smooth, nonsingular and involutive.

**Remark 2.2.6**
(i)     It follows from Frobenius' Theorem that for involutive distributions defined locally around a certain point the notions of leaf (of a foliation) and integral manifold (of a distribution) may be used interchangeably.
(ii)     If $V = \text{sp}\{e_1, \ldots, e_k\}$ is a subspace in $\mathbb{R}^n$ (with $\{e_1, \ldots, e_n\}$ denoting the standard basis for $\mathbb{R}^n$), then $V$ can be considered as an integral

manifold through $x = 0$ of the *flat* distribution $\Delta$ defined by $\Delta(x) = \text{sp}\{\frac{\partial}{\partial x_1}\big|_x, \ldots, \frac{\partial}{\partial x_k}\big|_x\}$ which will be denoted by $\Delta_V$.

(iii)   If $\Delta$ is a distribution on $\mathbb{R}^n$, then $\Delta(0)$ can be identified with a subspace in $\mathbb{R}^n$ (identifying $T_0\mathbb{R}^n$ with $\mathbb{R}^n$). The notation $\Delta(0)$ is used both for the subspace of $T_0\mathbb{R}^n$ and that of $\mathbb{R}^n$.                                                  □


A vector field $g$ on $\mathbb{R}^n$ is *tangent* to a manifold $S$ in $x \in S$ if $g(x) \in T_xS$. A vector field $g$ is *transversal* to a manifold $S$ in $\mathbb{R}^n$ if $g$ is not tangent to $S$ for all $x \in S$, i.e. $g(x) \notin T_xS$ for all $x \in S$. A constant dimensional distribution $\Delta$ is transversal to a manifold $S$ in $\mathbb{R}^n$ if for each $x \in S$, $\dim(\Delta(x) + T_xS) = \min\{d+m,n\}$, where $d = \dim \Delta$, $m = \dim S$. Hence, if $\Delta$ is transversal to $S$, then there are $\min\{d,n-m\}$ independent vector fields in $\Delta$ that are transversal to $S$. (Note that in general these vector fields may only be locally defined.)

A *codistribution* $\Omega$ on $\mathbb{R}^n$ is a rule assigning to each $x \in \mathbb{R}^n$ a subspace $\Omega(x) \subset T_x^*\mathbb{R}^n$ such that for every $x \in \mathbb{R}^n$, there exist a neighborhood $\mathcal{O}(x)$ of $x$ and a set of covector fields defined on $\mathcal{O}(x)$ denoted $\{\omega_i \mid i \in I\}$ with the property that $\Omega(y) = \text{sp}\{\omega_i(y) \mid i \in I\}$ for all $y \in \mathcal{O}(x)$. The codistribution is smooth if the covector fields $\{\omega_i \mid i \in I\}$ are.

If $\Delta$ is a distribution on $\mathbb{R}^n$ then the *annihilator* of $\Delta$ denoted as ann $\Delta$ is defined by


(2.2.13)   ann $\Delta(x) = \text{sp}\{\omega(x) \mid \omega$ covector field such that $\langle\omega,X\rangle = 0$

for all $X \in \Delta\}$

If $\Omega$ is a codistribution on $\mathbb{R}^n$ then the *kernel* of $\Omega$ denoted as ker $\Omega$ is defined by


(2.2.14)   ker $\Omega(x) = \text{sp}\{X(x) \mid X$ vector field such that $\langle\omega,X\rangle = 0$

for all $\omega \in \Omega\}$

The kernel and annihilator need not be smooth in general. However, if $\Delta$ and $\Omega$ are smooth and nonsingular, then so are ann $\Delta$ and ker $\Omega$ while ker ann $\Delta = \Delta$ and ann ker $\Omega = \Omega$ ([NvdS4]).


**Remark 2.2.7**   In the sequel the following short-hand notation is used.

(i)      If $\{h_1,\ldots,h_\ell\}$ is a set of smooth real-valued functions on $\mathbb{R}^n$ then $dh$ denotes the codistribution $\text{sp}\{dh_1,\ldots,dh_\ell\}$ (and ker $dh$ its kernel).

(ii)     If $\Delta$ is a distribution and $f$ is a vector field on $\mathbb{R}^n$ then $[f,\Delta]$ denotes the distribution spanned by the set of all Lie-brackets $[f,X]$ for any $X \in \Delta$, so $[f,\Delta] = \text{sp}\{[f,X] \mid X \in \Delta\}$.

(iii) We use abbreviations as $dx_1$ and $\frac{\partial}{\partial x_1}$ instead of $dx_1\big|_x$ and $\frac{\partial}{\partial x_1}\big|_x$ etc.$\square$

**Remark 2.2.8** If we identify $T_x\mathbb{R}^n$ with $\mathbb{R}^n$, then the vector field $f$ on $\mathbb{R}^n$ induces the differential equations $\dot{x} = f(x)$ on $\mathbb{R}^n$. In the sequel the vector field and differential equations point of view are used interchangeably. $\square$

**2.3 Controlled invariance and the Local Disturbance Decoupling Problem**

The solution of the Local Disturbance Decoupling Problem (LDDP) is by now well-known. In this section we introduce the concept of (local) controlled invariance and some related topics and we give the solution of the LDDP (following [Is] and [NvdS4]).
Recall that for linear systems the main idea behind the solution of the DDP is to find a controlled invariant subspace in the kernel of the output mapping that contains the disturbances and to make that subspace invariant. The idea of this section is to extend this linear paradigm to the nonlinear context and the first step in doing that is defining invariant and controlled invariant distributions. (Recall from Section 2.2 that a subspace may be considered as an integral manifold of a flat distribution.)

**Definition 2.3.1** A distribution $\Delta$ is said to be *invariant* under a vector field $f$ if $[f,\Delta] \subset \Delta$.

Recall from Section 2.2 that $[f,\Delta] \subset \Delta$ means that $[f,\tau] \in \Delta$ for all $\tau \in \Delta$.

**Remark 2.3.2** Let $V$ be a subspace in $\mathbb{R}^n$ and suppose that $V$ is invariant under the matrix $A$, i.e. $AV \subset V$. Let $\Delta_V$ denote the flat distribution that can be associated with $V$ and $f_A$ the vector field defined by $f_A(x) = Ax \in T_x\mathbb{R}^n$ for all $x \in \mathbb{R}^n$.
Suppose that $\{e_1,\ldots,e_k\}$ is a basis for $V$, then $\{\frac{\partial}{\partial x_1},\ldots,\frac{\partial}{\partial x_k}\}$ is a basis for $\Delta_V$. Moreover,

$$(2.3.1) \quad [f_A,\frac{\partial}{\partial x_i}](x) = \left(\frac{\partial}{\partial x}(\frac{\partial}{\partial x_i})f_A\right)(x) - \left(\frac{\partial f_A}{\partial x}\frac{\partial}{\partial x_i}\right)(x) = 0 - Ae_i =$$

$$= -Ae_i, \qquad i = 1,\ldots,k, \quad x \in \mathbb{R}^n$$

Since, by assumption $Ae_i \in \mathcal{V}$, we have that $[f_A, \frac{\partial}{\partial x_i}](x) \in \Delta_{\mathcal{V}}(x)$ for all $x \in \mathbb{R}^n$. Hence, $[f_A, \Delta_{\mathcal{V}}] \subset \Delta_{\mathcal{V}}$.

It follows from these calculations that invariance of a distribution under a vector field is a nonlinear generalization of the invariance of a subspace under a linear mapping. □

**Lemma 2.3.3** Let $\Delta$ be a regular distribution of dimension $d$. Suppose that $\Delta$ is invariant under the vector field $f$. Then for each point $x_0$ there exist a neighborhood $\mathcal{O}(x_0)$ of $x_0$ and a coordinate transformation $z = \varphi(x)$ defined on $\mathcal{O}(x_0)$, in which the vector field $f$ can be represented by a vector of the form

$$(2.3.2) \quad \bar{f}(z) = \begin{bmatrix} f_1(z_1, \ldots, z_d, z_{d+1}, \ldots, z_n) \\ \vdots \\ f_d(z_1, \ldots, z_d, z_{d+1}, \ldots, z_n) \\ f_{d+1}(z_{d+1}, \ldots, z_n) \\ \vdots \\ f_n(z_{d+1}, \ldots, z_n) \end{bmatrix}$$

The concept of invariance of a distribution under a vector field can also be considered from a geometric point of view (cf. the notion of invariant foliation in Section 2.5). Suppose that $\Delta$ is a regular distribution invariant under the vector field $f$. Let $x_0$ and $x_1$ be two points belonging to a maximal integral manifold $S_0$ of $\Delta$ (see Figure 2).

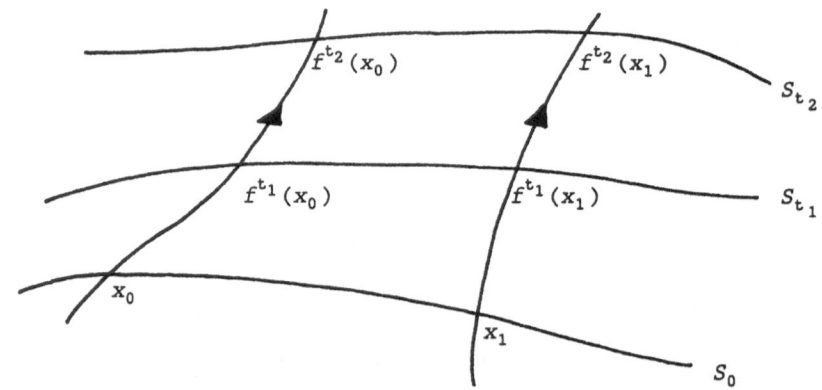

**Figure 2.** A geometric picture of invariance

Denote the solution of the differential equations $\dot{x} = f(x)$, $x(0) = \tilde{x}_0$ at time $t$ by $f^t(\tilde{x}_0)$, then $f^t(x_0)$ and $f^t(x_1)$ are contained in the same maximal integral manifold $S_t$ of $\Delta$. Sometimes we write this as $S_t = f^t(S_0)$. If for all $x_0$ the solution $f^t(x_0)$ exists for all $t$ then $f$ is said to be *complete*.

Consider the smooth nonlinear control system

$$(2.3.3) \quad \begin{cases} \dot{x} = f(x) + g(x)u, & x \in \mathbb{R}^n, \ u \in \mathbb{R}^m \\ y = h(x), & y \in \mathbb{R}^\ell \end{cases}$$

Recall from Chapter 1 that $\dot{x} = f(x) + g(x)u$ is short-hand notation for $\dot{x} = f(x) + \sum_{i=1}^{m} g_i(x)u_i$ and $y = h(x)$ for $y_i = h_i(x)$, $i = 1, \ldots, \ell$.

**Definition 2.3.4**

(i) A distribution is said to be *controlled invariant* on $\mathbb{R}^n$ if it is smooth, involutive and if there exists a smooth regular static state feedback

$$(2.3.4) \quad u = \alpha(x) + \beta(x)v, \quad u \in \mathbb{R}^m, \ v \in \mathbb{R}^m, \ x \in \mathbb{R}^n$$

defined on $\mathbb{R}^n$ such that $\Delta$ is invariant under the vector fields $\tilde{f} := f + g\alpha$ and $\tilde{g}_i := (g\beta)_i$, $i = 1, \ldots, m$, i.e.

$$(2.3.5) \quad [\tilde{f}, \Delta](x) \subset \Delta(x), \ [\tilde{g}_i, \Delta](x) \subset \Delta(x), \ i = 1, \ldots, m \quad \text{for all } x \in \mathbb{R}^n$$

(ii) A distribution $\Delta$ is said to be *locally controlled invariant* if for each $x \in \mathbb{R}^n$ there exists a neighborhood $\mathcal{O}(x)$ of $x$ with the property that $\Delta$ is controlled invariant on $\mathcal{O}(x)$.

(iii) A smooth regular static state feedback (2.3.4) that fulfills (2.3.5) is called a "friend" of $\Delta$, denoted by $(\alpha, \beta) \in \mathcal{F}(\Delta)$. If (2.3.5) holds, then $\Delta$ is said to be invariant under $\tilde{f}$ and $\tilde{g}$ (or under $\tilde{f}$ and $\tilde{g}_i$, $i = 1, \ldots, m$).

Let $G := \text{sp}\{g_1, \ldots, g_m\}$. The following lemma holds.

**Lemma 2.3.5** Let $\Delta$ be an involutive smooth distribution. Suppose that the distributions $\Delta$, $G$ and $\Delta + G$ are nonsingular on $\mathbb{R}^n$. Then $\Delta$ is locally controlled invariant if and only if

$(2.3.6a) \quad [f, \Delta] \subset \Delta + G$

$(2.3.6b) \quad [g_i, \Delta] \subset \Delta + G, \quad i = 1, \ldots, m$

**Remark 2.3.6** Note that in order that $\Delta$ is locally controlled invariant two conditions have to be fulfilled, while in order that a subspace is controlled invariant only one condition has to hold (see Definition 1.3(a)). Writing out (2.3.6) for a linear system with respect to the standard basis for $\mathbb{R}^n$ shows that the conditions $[f,\Delta] \subset \Delta + G$ and $A\mathcal{V} \subset \mathcal{V} + \text{im } B$ are equivalent, while $[g_i,\Delta] \subset \Delta + G$ automatically holds for the linear system. The extra condition (2.3.6b) is due to the fact that in general the vector fields $g_i$ depend explicitly on $x$, whereas the constant column vectors $b_i$ of $B$ do not. $\qquad \square$

The LDDP can be solved by making use of the concept of local controlled invariance. So, consider the following smooth nonlinear control system with disturbances

$$(2.3.7) \quad \begin{cases} \dot{x} = f(x) + g(x)u + p(x)q, & x \in \mathbb{R}^n, \ u \in \mathbb{R}^m, \ q \in \mathbb{R}^r \\ y = h(x), & y \in \mathbb{R}^\ell \end{cases}$$

where, as usual, $x$, $u$, $q$ and $y$ denote the states, inputs, disturbances and outputs.

**Lemma 2.3.7** There exists a unique largest involutive distribution $\Delta^*$ in $\text{ker } dh$ $(= \bigcap_{i=1}^{\ell} \text{ker } dh_i)$ that fulfills (2.3.6).

Note that, by definition, $\Delta^*$ is involutive. If the conditions of Lemma 2.3.5 are fulfilled for $\Delta^*$, then $\Delta^*$ is the largest locally controlled invariant distribution in $\text{ker } dh$. If so and if the disturbance vector fields are contained in $\Delta^*$, then the LDDP can be solved by applying an arbitrary smooth feedback $(\alpha,\beta) \in \mathcal{F}(\Delta^*)$. In fact, the following result holds.

**Theorem 2.3.8** Consider the smooth nonlinear system (2.3.7). Suppose that the distributions $\Delta^*$, $\Delta^* + G$ and $G$ are nonsingular. Then the Local Disturbance Decoupling Problem is solvable if and only if $\text{sp}\{p_1,\ldots,p_r\} \subset \Delta^*$.

The result in Theorem 2.3.8 is appealing, because there exists an algorithm that calculates $\Delta^*$ in case some regularity conditions hold.

**Algorithm 2.3.9** *Controlled Invariant Distribution Algorithm*

1.  $\Omega_0 := dh$

2.  $\Omega_k := \Omega_{k-1} + L_f(\Omega_{k-1} \cap \text{ann } G) + \sum_{i=1}^{m} L_{g_i}(\Omega_{k-1} \cap \text{ann } G), \quad k = 1,2,\ldots$

Here $L_f(\Omega_{k-1} \cap \text{ann } G)$ denotes the codistribution spanned by the covectors $L_f\omega$ with $\omega \in \Omega_{k-1} \cap \text{ann } G$.


**Lemma 2.3.10** Assume that the codistributions ann $G$, $\Omega_k$ and $\Omega_k \cap \text{ann } G$ have constant dimension for all $k \geq 0$. Then $\Delta^*$ is given by the kernel of $\Omega_n$, i.e. $\Delta^* = \ker \Omega_n$. Moreover, $\Delta^*$, $G$ and $\Delta^* + G$ are nonsingular.


If the codistributions $\Omega_k$ are nonsingular on a neighborhood of $x_0$, then $x_0$ is said to be a regular point of the Controlled Invariant Distribution Algorithm.

In many cases it is not necessary to use Algorithm 2.3.9 to calculate $\Delta^*$ (see Theorem 2.3.12).

**Definition 2.3.11** Consider the smooth system (2.3.3). The *relative degree* $r_i(x)$ ($i = 1,\ldots,\ell$) is the smallest integer such that

(2.3.8)
$$\begin{cases} L_{g_j}L_f^k h_i(x) = 0, & j = 1,\ldots,m, \ k < r_i(x)-1 \\[2mm] L_{g_j}L_f^{r_i(x)-1}h_i(x) \neq 0 & \text{for some } j \end{cases}$$

If the integers $r_1(x),\ldots,r_\ell(x)$ are finite and constant, say equal to $r_1,\ldots,r_\ell$, then the *decoupling matrix* $A(x)$ is defined by

(2.3.9) $\quad (A(x))_{ij} = L_{g_j}L_f^{r_i-1}h_i(x)$

**Theorem 2.3.12** Consider the smooth nonlinear system (2.3.3). Assume that the relative degrees $r_i(x)$, $i = 1,\ldots,\ell$ are constant and finite, say equal to $r_i$, $i = 1,\ldots,\ell$, on a neighborhood $\mathcal{O}$ of $x_0$, and that the decoupling matrix (2.3.9) has full row rank on $\mathcal{O}$. Then on $\mathcal{O}$, $\Delta^*$ is given by

(2.3.10) $\quad \Delta^* = \bigcap_{i=1}^{\ell} \bigcap_{k=0}^{r_i-1} \ker dL_f^k h_i$

Moreover, a regular static state feedback solving the LDDP follows from the equations

(2.3.11) $\quad A(x)\alpha(x) + b(x) = 0, \quad A(x)\beta(x) = \begin{pmatrix} I_\ell & 0 \end{pmatrix}$

where $b(x)$ is defined by $\left(b(x)\right)_i = L_f^{r_i} h_i(x)$, $i = 1,\ldots,\ell$.

**Definition 2.3.13** A point $x_0 \in \mathbb{R}^n$ is said to be a *regular point* of the system (2.3.3) if the relative degrees $r_i(x)$, $i = 1,\ldots,\ell$ are finite and constant on a neighborhood $\mathcal{O}$ of $x_0$ and $x_0$ is called a *regular point of the decoupling matrix* of (2.3.3) if $x_0$ is a regular point and if the decoupling matrix has full row rank on $\mathcal{O}$.

The term decoupling matrix refers to the fact that this matrix plays a crucial role in the (Strong) Input–Output Decoupling Problem (also referred to as the Noninteracting Control Problem, see e.g. [SR], [IKGM1], [Fr]). The problem formulation given below is taken from [NvdS4]. We restrict ourselves here to *square* systems, i.e. systems with an equal number of inputs and outputs and we briefly comment on the formulation for nonsquare systems. Extensions can be found in Section 6.1 and in the literature (see e.g. [NvdS4] and references therein).

**Definition 2.3.14** Consider the square smooth system (2.3.3) and a point $x_0 \in \mathbb{R}^n$. This system is said to be *strongly input–output decoupled* around $x_0$ if there exists a neighborhood $\mathcal{O}$ of $x_0$ for which

$$(2.3.12) \quad L_{g_j} L_{X_k} \cdots L_{X_1} h_i(x) = 0, \quad \forall k \geq 0, \quad X_1,\ldots,X_k \in \{f, g_1,\ldots,g_m\}$$

for all $x$ in $\mathcal{O}$ and $i,j = 1,\ldots,m$, $i \neq j$, and if the relative degrees $r_1(x),\ldots,r_m(x)$ are finite and constant, say equal to $r_1,\ldots,r_m$, on $\mathcal{O}$.

Note that if the system is strongly input–output decoupled, then the set

$$(2.3.13) \quad S = \{x \in \mathbb{R}^n \mid L_{g_i} L_f^{r_i-1} h_i(x) \neq 0, \ i = 1,\ldots,m\}$$

contains $\mathcal{O}$.

We can define the Strong Local Input–Output Decoupling Problem now.

**Definition 2.3.15** *Strong Local Input–Output Decoupling Problem* (SLIODP) Consider the square smooth nonlinear system (2.3.3) and a point $x_0 \in \mathbb{R}^n$. Under what conditions can we find a smooth regular static state feedback (2.3.4) defined on a neighborhood $\mathcal{O}$ of $x_0$ such that the feedback system (2.3.3,4) is strongly input–output decoupled around $x_0$?

25

**Theorem 2.3.16** Consider the square smooth nonlinear system (2.3.3) and a point $x_0 \in \mathbb{R}^n$. Assume that the system has finite and constant relative degrees $r_1, \ldots, r_m$ in a neighborhood of $x_0$. Then the Strong Local Input-Output Decoupling Problem is solvable around $x_0$ if and only if $x_0$ is a regular point of the decoupling matrix.

**Remark 2.3.17** A nonsquare system with $m$ inputs, $\ell$ outputs, $\ell < m$ is called strongly input-output decoupled around $x_0$ if there exists a neighborhood $\mathcal{O}$ of $x_0$ for which (2.3.12) holds for all $x$ in $\mathcal{O}$ and $i = 1, \ldots, \ell$, $j = 1, \ldots, m$, $i \neq j$ and the relative degrees $r_1(x), \ldots, r_\ell(x)$ are finite and constant on $\mathcal{O}$. Then for nonsquare systems having constant, finite relative degrees on $\mathcal{O}$ Theorem 2.3.16 is also valid, i.e. the SLIODP is solvable around $x_0$ if and only if the decoupling matrix has full row rank. $\qquad$ $\square$

A control system can very well be input-output decoupled without being strongly input-output decoupled as follows from the following example.

**Example 2.3.18** ([NvdS4]) Consider the system (2.3.3) with $n = 2$, $m = \ell = 1$ and

$$(2.3.14) \quad f(x) = \begin{pmatrix} x_2^3 \\ 0 \end{pmatrix}, \qquad g(x) = \begin{pmatrix} 0 \\ 1 \end{pmatrix}, \qquad h(x) = x_1$$

Now $L_g h(x) = 0$, $L_g L_f h(x) = 3x_2^2$, so $r(x) = 2$ if $x_2 \neq 0$. Moreover, $L_f^2 h(x) = 0$, so $r(x) = \infty$ if $x_2 = 0$. Obviously, the conditions for strong input-output decouplability around $x_0 = 0$ are not fulfilled. However,

$$(2.3.15) \quad y^{(4)} = 6u^3 + 12x_2 u\dot{u} + 3x_2^2 u^{(2)}$$

so the output is indeed influenced by the input, whatever initial condition $x_0$ is chosen. $\qquad$ $\square$

An important role in the next chapters is played by the largest local controllability distribution in ker $dh$. This distribution, which exists in case some regularity conditions hold, will be denoted by $\Pi^*$.

**Definition 2.3.19** A distribution $\Delta$ on $\mathbb{R}^n$ is said to be a *controllability distribution* on $\mathbb{R}^n$ if it is smooth, involutive and there exists a regular static state feedback (2.3.4) defined on $\mathbb{R}^n$ and a subset $I \subset \{1, \ldots, m\}$ with the property that $\Delta \cap G = \mathrm{sp}\{(g\beta)_i \mid i \in I\}$ and $\Delta$ is the smallest distribution which is invariant under the vector fields $f + g\alpha$ and $(g\beta)_i$, $i = 1, \ldots, m$ and contains $(g\beta)_i$ for all $i \in I$.

$\Delta$ is called a *local controllability distribution* defined around $x_0$ if the feedback is defined in a neighborhood of $x_0$.

**Algorithm 2.3.20** *Controllability Distribution Algorithm*

1. $\quad \Delta_0 := \Delta^* \cap G$

2. $\quad \Delta_k := \Delta^* \cap \left( [f, \Delta_{k-1}] + \sum_{i=1}^{m} [g_i, \Delta_{k-1}] + G \right), \quad k = 1, 2, \ldots$

**Lemma 2.3.21** ([Is]) Consider the smooth nonlinear system (2.3.3). Assume that $\Delta^*$, $G$ and $\Delta^* + G$ are nonsingular. Suppose that the Controllability Distribution Algorithm ends in $\kappa^*$ steps and that $\Delta_{\kappa^*}$ is nonsingular. Then the largest local controllability distribution $\Pi^*$ in $\ker dh$ exists and $\Pi^*$ equals $\Delta_{\kappa^*}$.

In the sequel the following properties of $\Pi^*$ are used.

(i)  If the conditions of Lemma 2.3.21 hold, and if $(\alpha, \beta) \in \mathscr{F}(\Delta^*)$, then $(\alpha, \beta) \in \mathscr{F}(\Pi^*)$ (see [Is]).

(ii) If $x_0$ is a regular point of the decoupling matrix for a square system, then $\Pi^* = 0$ (see [Nij1]).

**Remark 2.3.22** Note that $\Pi^*$ is a possible nonlinear analogue for $\mathscr{R}^*$, the largest controllability subspace in the kernel of the output mapping. It is well-known (see [Wo]) that for linear systems the dynamics restricted to $\mathscr{R}^*$ are controllable (so, in particular, stabilizable). For nonlinear systems there is no direct relation between controllability distributions and stabilizability. In fact, in the following example it is shown that the dynamics of a system restricted to the leaf of a controllability distribution through an equilibrium (see Section 2.5) need not be stabilizable. □

**Example 2.3.23** Consider the system (2.3.3) with $n = 5$, $m = 2$, $\ell = 1$ and

(2.3.16) $\quad f(x) = x_4 \dfrac{\partial}{\partial x_4}, \quad g_1(x) = \dfrac{\partial}{\partial x_2}, \quad g_2(x) = x_2 \dfrac{\partial}{\partial x_1} + (1 + x_1) \dfrac{\partial}{\partial x_4} + \dfrac{\partial}{\partial x_5},$

$$h(x) = x_5$$

Since $<dh, g_2>(x) = 1$ for all $x$, we have that $\Delta^* = \text{sp}\{\dfrac{\partial}{\partial x_1}, \dfrac{\partial}{\partial x_2}, \dfrac{\partial}{\partial x_3}, \dfrac{\partial}{\partial x_4}\}$. From

Algorithm 2.3.20 it follows that $\Pi^* = \text{sp}\{\frac{\partial}{\partial x_1}, \frac{\partial}{\partial x_2}, \frac{\partial}{\partial x_4}\}$. Now the dynamics of this system restricted to the leaf $L_0$ of $\Pi^*$ through $x = 0$ are given by

$$(2.3.17) \quad \dot{x}_1 = 0, \qquad \dot{x}_2 = u_1, \qquad \dot{x}_4 = x_4$$

Clearly, these dynamics are unstable, whatever $u_1$ is. □

**Remark 2.3.24** In the literature the concept of controlled invariance is defined also for more general nonlinear systems. In [NvdS1] controlled invariance is defined for systems of the form

$$(2.3.18) \quad \dot{x} = f(x,u)$$

The Disturbance Decoupling Problem for nonlinear systems

$$(2.3.19) \quad \begin{cases} \dot{x} = f(x,u,d) \\ y = h(x,u) \end{cases}$$

(where also the disturbances enter in a nonlinear way) is treated in [NvdS2], see also [NvdS4]. □

**2.4 Problem formulation and quick overview of Sections 2.1, 2.2 and 2.3**

In this section the problem formulation of the Local Disturbance Decoupling Problem with Stability is recapitulated. Moreover, the most important results from the previous sections are listed here and some conventions are introduced.

Consider the smooth nonlinear control system

$$(2.4.1) \quad \begin{cases} \dot{x} = f(x) + g(x)u + p(x)q, & x \in \mathbb{R}^n, \ u \in \mathbb{R}^m, \ q \in \mathbb{R}^r \\ y = h(x), & y \in \mathbb{R}^\ell \end{cases}$$

where $x$, $u$, $q$ and $y$ denote the states, inputs, disturbances and outputs, respectively. A feedback

$$(2.4.2) \quad u = \alpha(x) + \beta(x)v, \qquad u \in \mathbb{R}^m, \ v \in \mathbb{R}^m, \ x \in \mathbb{R}^n$$

is called a regular static state feedback if the matrix $\beta(x)$ is nonsingular for all $x$.

**Definition 2.4.1** *Local Disturbance Decoupling Problem with Stability* (LDDPS) Consider the smooth nonlinear system (2.4.1) with $f(0) = 0$. Under what conditions can we find a smooth regular static state feedback (2.4.2) defined locally around $x = 0$ with $\alpha(0) = 0$ such that in the feedback system (2.4.1,2) the disturbances $q$ do not influence the outputs $y$, and $x = 0$ is a locally exponentially stable equilibrium of the modified drift dynamics $\dot{x} = f(x)+g(x)\alpha(x)$?

In the sequel we use the following notations:

$G = \text{sp}\{g_1,\ldots,g_m\}$

$P = \text{sp}\{p_1,\ldots p_r\}$

$\text{ker } dh = \overset{\ell}{\underset{i=1}{\cap}} \text{ker } dh_i$

$\Delta^*$: the largest locally controlled invariant distribution in ker $dh$

$\Pi^*$: the largest local controllability distribution in ker $dh$

$(\alpha,\beta) \in \mathcal{F}(\Delta^*)$: $[f+g\alpha,\Delta^*] \subset \Delta^*$, $[(g\beta)_i,\Delta^*] \subset \Delta^*$, $\quad i = 1,\ldots,m$
on a neighborhood of $x = 0$ (In words: "$\Delta^*$ is invariant under $f + g\alpha$ and $(g\beta)_i$, $i = 1,\ldots,m$ or, in short, $\Delta^*$ is invariant under $f + g\alpha$ and $g\beta$").

The relative degree $r_i(x)$ $(i = 1,\ldots,\ell)$ is the smallest integer such that

$$(2.4.3) \quad \begin{cases} L_{g_j}L_f^k h_i(x) = 0, & j = 1,\ldots,m, \ k < r_i(x)-1 \\ \\ L_{g_j}L_f^{r_i(x)-1}h_i(x) \neq 0 & \text{for some } j \end{cases}$$

If the integers $r_1(x),\ldots,r_\ell(x)$ are finite and constant, say equal to $r_1,\ldots,r_\ell$, the decoupling matrix $A(x)$ is defined by

$$(2.4.4) \quad \bigl(A(x)\bigr)_{ij} = L_{g_j}L_f^{r_i-1}h_i(x)$$

A point $x_0 \in \mathbb{R}^n$ is said to be a regular point of the smooth system (2.4.1) if the relative degrees $r_1(x),\ldots,r_\ell(x)$ are finite and constant on a neighborhood $\mathcal{O}$ of $x_0$ and $x_0$ is called a regular point of the decoupling matrix of (2.4.1) if $x_0$ is a regular point and if the decoupling matrix has full row rank on $\mathcal{O}$.

Assume that

(A1)  $x_0$ is a regular point of the decoupling matrix.

Then on $\mathcal{O}$, $\Delta^*$ is given by

$$(2.4.5) \quad \Delta^* = \bigcap_{i=1}^{\ell} \bigcap_{k=0}^{r_i-1} \ker dL_f^k h_i$$

Moreover, $\Delta^*$ and $\Delta^* + G$ are nonsingular. If the system (2.4.1) is square, i.e. $\ell = m$, then also $G$ is nonsingular. Furthermore, a special regular static state feedback $(\alpha, \beta) \in \mathcal{F}(\Delta^*)$ follows from

$$(2.4.6) \quad A(x)\alpha(x) + b(x) = 0, \quad A(x)\beta(x) = \begin{pmatrix} I_\ell & 0 \end{pmatrix}$$

where $b(x)$ is defined by $\big(b(x)\big)_i = L_f^{r_i} h_i(x)$, $i = 1, \ldots, \ell$.

If (A1) holds and if the Controllability Distribution Algorithm 2.3.20 ends in a finite number of steps, say $\kappa^*$, and $\Delta_{\kappa^*}$ is nonsingular, then $\Pi^*$ is nonsingular. Moreover, if $(\alpha, \beta) \in \mathcal{F}(\Delta^*)$, then $(\alpha, \beta) \in \mathcal{F}(\Pi^*)$. If (A1) holds for a square system, then $\Pi^* = 0$.

In the sequel all vector fields, functions, codistributions etc. are assumed to be smooth, unless stated otherwise.


## 2.5 Constrained, zero and restricted zero dynamics


We start this section by defining two concepts that are related to the invariance of a distribution and the notion of (strong) accessibility. Furthermore, we recall the definitions of constrained and zero dynamics and we introduce the new concept of restricted zero dynamics for nonlinear systems. Finally, we summarize some results from the theory of dynamical systems and we define what exponentially minimum phase nonlinear systems are.

Consider the smooth system

$$(2.5.1) \quad \begin{cases} \dot{x} = f(x) + g(x)u, \quad f(0) = 0, \ x \in \mathbb{R}^n, \ u \in \mathbb{R}^m \\ y = h(x), \qquad\qquad h(0) = 0, \ y \in \mathbb{R}^\ell \end{cases}$$

Let $\Delta$ denote a locally controlled invariant $k$-dimensional distribution contained in $\Delta^*$. Suppose that $\Delta \cap G$ is constant dimensional. Choose a regular static state feedback

(2.5.2)    $u = \alpha(x) + \beta(x)v,$    $u \in \mathbb{R}^m, v \in \mathbb{R}^m, x \in \mathbb{R}^n$

with $\alpha(0) = 0$ and $(\alpha,\beta) \in \mathcal{F}(\Delta)$. Without loss of generality, we may assume that $\Delta = \text{sp}\{\frac{\partial}{\partial x_1}\}$, $G \cap \Delta = \text{sp}\{\tilde{g}_1,\ldots,\tilde{g}_s\}$ and that the system (2.5.1,2) has the form

(2.5.3)    $\begin{cases} \dot{x}_1 = \tilde{f}^1(x_1,x_2) + \tilde{g}^{11}(x_1,x_2)u^1 + \tilde{g}^{12}(x_1,x_2)u^2 \\ \dot{x}_2 = \tilde{f}^2(x_2) \qquad\qquad\qquad + \tilde{g}^{22}(x_2)u^2 \\ y = h(x_2) \end{cases}$

where $\tilde{g}^1(x) = (\tilde{g}_1,\ldots,\tilde{g}_s)(x) = (\tilde{g}^{11}(x_1,x_2)^T \ \ 0^T)^T$ etc. The *dynamics of the system (2.5.1) restricted to the leaf $S_0$ of $\Delta$ through $x = 0$* are defined as the dynamics

(2.5.4)    $\dot{x}_1 = \tilde{f}^1(x_1,0) + \tilde{g}^{11}(x_1,0)u^1$

Note that $S_0 = \{x|\ x_2 = 0\}$ and that these dynamics are obtained by choosing $u^2 = 0$ in (2.5.3), which by $\tilde{f}(0) = 0$ implies that $\dot{x}_2 = 0$. The *dynamics of the system (2.5.1) modulo $\Delta$* are given by

(2.5.5)    $\dot{\bar{x}}_2 = \tilde{f}^2(\bar{x}_2) + \tilde{g}^{22}(\bar{x}_2)u^2$

The concepts that are introduced next in this section are related to controllability of a nonlinear system (see e.g. [Kr1], [SJ]). Consider the smooth nonlinear system

(2.5.6)    $\dot{x} = f(x) + g(x)u,$    $x \in \mathbb{R}^n, u \in \mathbb{R}^m$

Let, for any neighborhood $\mathcal{O}$ of $x_0$ and $T > 0$, $\mathcal{R}(x_0,\mathcal{O},T)$ denote the set of points that can be reached from $x_0$ (by applying piecewise constant controls) following trajectories which remain for $0 \leq t \leq T$ in $\mathcal{O}$. Let

(2.5.7)    $\mathcal{R}_T(x_0,\mathcal{O}) = \bigcup_{\tau \leq T} \mathcal{R}(x_0,\mathcal{O},\tau)$

The system (2.5.6) is called *locally accessible from $x_0$* if $\mathcal{R}_T(x_0,\mathcal{O})$ contains a nonempty open set in $\mathbb{R}^n$ for all neighborhoods $\mathcal{O}$ of $x_0$ and all $T > 0$. The system (2.5.6) is locally accessible from $x_0$ if dim $C(x_0) = n$, where $C$ is the distribution generated by

(2.5.8)    $\mathcal{C} := \{f, g_1, \ldots, g_m\}_{LA}$

the smallest Lie-algebra of vector fields containing $f$ and $g_1, \ldots, g_m$.
The system (2.5.6) is said to be *locally strongly accessible from* $x_0$ if for
each neighborhood $\mathcal{O}$ of $x_0$ the set $\mathcal{R}(x_0, \mathcal{O}, T)$ contains a nonempty open set in
$\mathbb{R}^n$ for any $T > 0$ sufficiently small. Let

(2.5.9)    $\mathcal{C}_0 = \{\mathrm{ad}_f^k g_i, \ i = 1, \ldots, m, \ k \geq 0\}_{LA}$

denote the smallest Lie-algebra containing the vector fields $\mathrm{ad}_f^k g_i$,
$i = 1, \ldots, m, \ k \geq 0$. The system (2.5.6) is locally strongly accessible from
$x_0$ if dim $C_0(x_0) = n$, where $C_0$ is the distribution generated by $\mathcal{C}_0$.
Note that if $x_0$ is an equilibrium of $f$, i.e. $f(x_0) = 0$, then $C_0(x_0)$ and
$C(x_0)$ coincide. The distribution $C$ is called the accessibility distribution
and $C_0$ the strong accessibility distribution.


During the last decade quite some attention has been paid to finding the
nonlinear equivalent of transmission zeros for linear systems (see e.g.
[KI], [IM]). This led to the definition of constrained and zero dynamics
(manifolds) (see e.g. [IM], [vdS2]). As shown in [BI3] the zero dynamics
play an important role in the stabilization of minimum phase systems.
However, it turns out that for the solution of the LDDPS the dynamics
restricted to a submanifold of the zero dynamics manifold (to which we will
refer as restricted zero dynamics) play an important role. For convenience
we recall the definitions of constrained and zero dynamics first.
Consider again the smooth nonlinear system (2.5.1). The *constrained* (or
*clamped* or *zero-output constrained*) dynamics of the system (2.5.1) are the
dynamics of this system compatible with the constraints $y = 0$. These
dynamics can be calculated using a modified version of Hirschorn's
structure algorithm ([Hi1]) or Krener's algorithm ([Kr2]). Geometrically,
the latter can be formulated as follows (see [IM], [Is], [NvdS4], [vdS2]).
(As usual, $\mathcal{O}$ denotes a neighborhood of $x_0 = 0$.)


**Algorithm 2.5.1**

1.   Define $N_0 := h^{-1}(0)$

2.   Assume that the manifolds $N_0, \ldots, N_{k-1}$ through $x_0 = 0$ are defined.
     Suppose that $N_{k-1} \cap \mathcal{O}$ is a smooth manifold and let $N_{k-1}^c$ denote the
     connected component of $N_{k-1} \cap \mathcal{O}$ containing $x_0 = 0$. Define $N_k$ by
     $$N_k := \{x \in N_{k-1}^c \mid f(x) \in T_x N_{k-1}^c + \mathrm{sp}\{g_1(x), \ldots, g_m(x)\}\}$$

We say that $x_0 = 0$ is a regular point for Algorithm 2.5.1 if we can find a neighborhood $\mathcal{O}$ such that for every $k$ $N_k \cap \mathcal{O}$ is a smooth manifold through $x_0 = 0$. In case $x_0 = 0$ is a regular point the algorithm converges in $k^* < n$ steps. Let $N^*$ denote the connected component of $N_{k^*} \cap \mathcal{O}$ through $x_0 = 0$. Suppose that dim $G = m$ and that for all $x \in N^*$ $G(x) \cap T_x N^*$ has constant dimension. Then $N^*$ is called the *constrained dynamics manifold*. Without loss of generality, we may assume that $N^* = \{x| \ x_2 = 0\}$. Then the constrained dynamics are given by

$$(2.5.10) \quad \dot{x}_1 = \tilde{f}^1(x_1,0) + \tilde{g}^1(x_1,0)v$$

with inputs $v$. In general, the dimension of $v$ is strictly less than the dimension of $u$. Assume that the strong accessibility distribution $\tilde{C}_0$ of the system (2.5.10) has constant dimension on $\mathcal{O}$, then locally around $x_0 = 0$ $N^*$ may be factored out by the integral manifolds of the distribution $\tilde{C}_0$, denoted by $N^*/\tilde{C}_0$. The restriction of the constrained dynamics (2.5.10) to $N^*/\tilde{C}_0$ is called the *zero dynamics* of the system (2.5.1) (cf. [vdS2]).

Before we define the restricted zero dynamics for the general system (2.5.1), we define this concept for strongly input-output decouplable systems. So, assume for the moment that (A1) holds (see Section 2.4). Recall from the previous section that this means that the relative degrees $r_1,\ldots,r_\ell$ are constant and that the decoupling matrix $A(x)$ has full row rank on $\mathcal{O}$. Assume, furthermore, that $\Pi^*$ has constant dimension on $\mathcal{O}$. In case $\mathcal{O}$ is small enough, there exist a smooth regular static state feedback (2.5.2) with $\alpha(0) = 0$ and $(\alpha,\beta) \in \mathcal{F}(\Delta^*)$ and a coordinate transformation $z = \varphi(x)$ defined on $\mathcal{O}$ such that, with $\tilde{f} := f + g\alpha$ and $\tilde{g}_i := (g\beta)_i$ $(i = 1,\ldots,m)$,

$$(2.5.11) \quad \Delta^* \cap G = \mathrm{sp}\{\tilde{g}_1,\ldots,\tilde{g}_{m-\ell}\}, \quad \Pi^* = \mathrm{sp}\{\frac{\partial}{\partial z_1}\}, \quad \Delta^* = \mathrm{sp}\{\frac{\partial}{\partial z_1},\frac{\partial}{\partial z_2}\}$$

and

$$(2.5.12) \quad z_{i+2} = (h_i \ L_{\tilde{f}}h_i \ \cdots \ L_{\tilde{f}}^{r_i-1}h_i)^{\mathrm{T}}, \quad i = 1,\ldots,\ell$$

Then the system (2.5.1,2) has the form

$$(2.5.13) \quad \begin{cases} \dot{z}_1 &= \hat{f}^1(z_1,\ldots,z_{\ell+2}) + \hat{g}^{11}(z_1,\ldots,z_{\ell+2})v^1 + \hat{g}^{12}(z_1,\ldots,z_{\ell+2})v^2 \\ \dot{z}_2 &= \hat{f}^2(z_2,\ldots,z_{\ell+2}) \hspace{5em} + \hat{g}^{22}(z_2,\ldots,z_{\ell+2})v^2 \\ \dot{z}_{i+2} &= A^{i+2}z_{i+2} \hspace{9em} + B^{i+2}v_{m-\ell+i}, \\ y_i &= z_{i+2}^1 \hspace{13em} i = 1,\ldots,\ell \end{cases}$$

where

$$(2.5.14) \quad A^{i+2} = \begin{bmatrix} 0 & 1 & & 0 \\ & \ddots & \ddots & \\ & & & 1 \\ 0 & & & 0 \end{bmatrix}_{r_i \times r_i} , \qquad B^{i+2} = \begin{bmatrix} 0 \\ \vdots \\ 0 \\ 1 \end{bmatrix}_{r_i} , \qquad i = 1, \ldots, \ell$$

$$(2.5.15) \quad v^1 = (v_1, \ldots, v_{m-\ell}), \qquad v^2 = (v_{m-\ell+1}, \ldots, v_m)$$

and

$$(2.5.16a) \quad \tilde{f}(z) = \begin{bmatrix} \hat{f}^1(z_1, \ldots, z_{\ell+2}) \\ \hat{f}^2(z_2, \ldots, z_{\ell+2}) \\ A^3 z_3 \\ \vdots \\ A^{\ell+2} z_{\ell+2} \end{bmatrix}$$

$$(2.5.16b) \quad \tilde{g}^1(z) = (\tilde{g}_1, \ldots, \tilde{g}_{m-\ell})(z) = \begin{bmatrix} \hat{g}^{11}(z_1, \ldots, z_{\ell+2}) \\ 0 \\ \vdots \\ 0 \end{bmatrix}$$

$$(2.5.16c) \quad \tilde{g}^2(z) = (\tilde{g}_{m-\ell+1}, \ldots, \tilde{g}_m)(z) = \begin{bmatrix} \hat{g}^{12}(z_1, \ldots, z_{\ell+2}) \\ \hat{g}^{22}(z_2, \ldots, z_{\ell+2}) \\ \begin{bmatrix} B^3 & & 0 \\ & \ddots & \\ 0 & & B^{\ell+2} \end{bmatrix} \end{bmatrix}$$

The constrained dynamics of the system (2.5.1) can be obtained as follows. In order that the outputs $y(t)$ are identically equal to zero it is necessary that the inputs $v_{m-\ell+i}$, $i = 1, \ldots, \ell$ in (2.5.13) are chosen equal to zero. Hence, necessary for the outputs $y$ to be identically equal to zero is that $z_i = 0$, $i = 3, \ldots, \ell+2$. The constrained dynamics manifold is given by

$$(2.5.17) \quad M_0 := \{ z \in \mathcal{O} \,|\, z_3 = z_4 = \cdots = z_{\ell+2} = 0 \}$$

The dynamics on this locally defined manifold $M_0$ are given by

$$(2.5.18) \quad \begin{cases} \dot{z}_1 = \hat{f}^1(z_1, z_2, 0, \ldots, 0) + \hat{g}^{11}(z_1, z_2, 0, \ldots, 0)v^1 \\ \dot{z}_2 = \hat{f}^2(z_2, 0, \ldots, 0) \end{cases}$$

Note that the manifold $M_0$ is exactly the leaf of the distribution $\Delta^*$ through $x_0 = 0$ and that the dynamics (2.5.18) can also be obtained by restricting the dynamics of the system (2.5.1) to this manifold $M_0$. As noted earlier, the zero dynamics can be found now by factoring out $M_0$ by the leaves of the accessibility distribution $\tilde{C}_0$ of (2.5.18). Let $\tilde{\Pi}^* := P\Pi^*$, where $P$ denotes the canonical projection from $\mathcal{O}$ onto $M_0$. It is also possible to factor out the leaves of the distribution $\tilde{\Pi}^*$. The restriction of the constrained dynamics (2.5.18) to $M_0/\tilde{\Pi}^*$ is said to be the *restricted zero dynamics* and is given by

$$(2.5.19) \quad \dot{\bar{z}}_2 = \hat{f}^2(\bar{z}_2, 0, \ldots, 0)$$

Since $\tilde{C}_0 \subset \tilde{\Pi}^*$ it is obvious that the restricted zero dynamics manifold is contained in the zero dynamics manifold. If the system (2.5.1) is square, the distribution $\Pi^* = 0$ and it is easily seen that in that case the constrained, zero and restricted zero dynamics coincide. In general, the zero dynamics and the restricted zero dynamics are not the same, as can be seen from the following example.

**Example 2.5.2** (cf. Example 2.3.23) Consider the system

$$(2.5.20) \quad \dot{x} = f(x) + g_1(x)u_1 + g_2(x)u_2, \quad y = h(x), \quad x \in \mathbb{R}^5$$

with

$$(2.5.21) \quad f(x) = x_4 \frac{\partial}{\partial x_4}, \quad g_1(x) = \frac{\partial}{\partial x_2}, \quad g_2(x) = x_2 \frac{\partial}{\partial x_1} + (1+x_1)\frac{\partial}{\partial x_4} + \frac{\partial}{\partial x_5},$$
$$h(x) = x_5$$

Since $\Delta^* = \text{sp}\{\frac{\partial}{\partial x_1}, \frac{\partial}{\partial x_2}, \frac{\partial}{\partial x_3}, \frac{\partial}{\partial x_4}\}$ the constrained dynamics manifold is $M_0 = \{x \in \mathbb{R}^5 | x_5 = 0\}$ and the constrained dynamics are given by

$$(2.5.22) \quad \dot{x}_1 = 0, \quad \dot{x}_2 = u_1, \quad \dot{x}_3 = 0, \quad \dot{x}_4 = x_4$$

The strong accessibility distribution of the system (2.5.22) is given by $\text{sp}\{\frac{\partial}{\partial x_2}\}$. Hence, the zero dynamics are given by

(2.5.23) $\dot{\bar{x}}_1 = 0$, $\quad \dot{\bar{x}}_3 = 0$, $\quad \dot{\bar{x}}_4 = \bar{x}_4$

On the other hand, since $\Pi^* = \text{sp}\{\frac{\partial}{\partial x_1},\frac{\partial}{\partial x_2},\frac{\partial}{\partial x_4}\}$ the restricted zero dynamics are given by

(2.5.24) $\dot{\bar{x}}_3 = 0$ $\qquad\qquad\qquad\qquad\qquad\qquad\qquad$ □

We return to the general case now. From now on, assumption (A1) does no longer hold true for the system (2.5.1). Assume that the distributions $\Delta^*$, $G$ and $\Delta^* + G$ have dimensions $k$, $m$ and $k+p$, respectively. It can easily be seen from Algorithm 2.5.1 that the integral manifold $M_0$ of $\Delta^*$ through $x_0 = 0$ is contained in the constrained dynamics manifold $N^*$. For, choose a regular static state feedback (2.5.2) with $\alpha(0) = 0$ and $(\alpha,\beta) \in \mathcal{F}(\Delta^*)$. Since $(f+g\alpha)(0) = 0$, it is obvious that the vector field $f+g\alpha$ is tangent to $M_0$. Now $M_0$ is contained in $h^{-1}(0)$ and for all $x$ in $M_0$ we have that $f(x)+g(x)\alpha(x) \in T_x M_0$, so $f(x) \in \text{sp}\{g_1(x),\ldots,g_m(x)\} + T_x M_0$. This implies that $M_0 \subset N_k$ for all $k$ and, in case $x_0$ is a regular point of Algorithm 2.5.1, $M_0 \subset N^*$ (see also [IM]). If, moreover, $\Pi^*$ is a regular distribution, then there exists a coordinate transformation $z = \varphi(x)$ such that

(2.5.25) $\Delta^* \cap G = \text{sp}\{\tilde{g}_1,\ldots,\tilde{g}_{m-p}\}$, $\quad \Pi^* = \text{sp}\{\frac{\partial}{\partial z_1}\}$, $\quad \Delta^* = \text{sp}\{\frac{\partial}{\partial z_1},\frac{\partial}{\partial z_2}\}$

(2.5.26) $\begin{cases} \dot{z}_1 = \hat{f}^1(z_1,z_2,z_3) + \hat{g}^{11}(z_1,z_2,z_3)v^1 + \hat{g}^{12}(z_1,z_2,z_3)v^2 \\ \dot{z}_2 = \hat{f}^2(z_2,z_3) \qquad\qquad\qquad\quad + \hat{g}^{22}(z_2,z_3)v^2 \\ \dot{z}_3 = \hat{f}^3(z_3) \qquad\qquad\qquad\qquad\;\; + \hat{g}^{32}(z_3)v^2 \\ y = h(z_3) \end{cases}$

where $v^1 = (v_1,\ldots,v_{m-p})$, $v^2 = (v_{m-p+1},\ldots,v_m)$. Clearly, $M_0 = \{z \mid z_3 = 0\}$. By choosing $v^2 \equiv 0$ it follows that $z_3(t) = 0$ and thus $y(t) = 0$ for all $t$. This implies that the dynamics

(2.5.27) $\begin{cases} \dot{z}_1 = \hat{f}^1(z_1,z_2,0) + \hat{g}^{11}(z_1,z_2,0)v^1 \\ \dot{z}_2 = \hat{f}^2(z_2,0) \end{cases}$

are part (!) of the constrained dynamics. The restricted zero dynamics are now obtained by factoring out the leaves of the distribution $P\Pi^*$. Hence, these dynamics are given by

(2.5.28)  $\bar{z}_2 = \hat{f}^2(\bar{z}_2,0)$

It follows from the foregoing that in this general case the restricted zero dynamics are restricted in two ways. First, the dynamics are restricted to $M_0$ which yields part of the constrained dynamics (in general) and then $P\Pi^*$ rather than the accessibility distribution is factored out. It can be proved that the restricted zero dynamics are uniquely defined by factoring out the state space $\mathbb{R}^n$ by the leaves of $\Pi^*$ and application of Lemma 3.4 in [IG].

We proceed with some notions from the theory of dynamical systems now. Consider the differential equations

(2.5.29)  $\dot{x} = f(x), \quad x \in \mathbb{R}^n$

where $f$ is a smooth complete vector field with equilibrium $x = 0$. A manifold $S$ in $\mathbb{R}^n$ is called *invariant* under $f$ if the following holds: if $x_0 \in S$ arbitrary, then the solution of (2.5.29) with $x(0) = x_0$, denoted by $f^t(x_0)$, is contained in $S$ for all $t$. We sometimes write $f^t(S) \subset S$ if $S$ is invariant under $f$. $S$ is called a *stable (unstable) invariant* manifold if $S$ is invariant under $f$ and $f^t(x_0) \to 0$ if $t \to \infty$ ($t \to -\infty$) for all $x_0 \in S$. Now consider the linearization of (2.5.29) around the equilibrium $x = 0$

(2.5.30)  $\dot{z} = \frac{\partial f}{\partial x}(0)z$

The equilibrium $x = 0$ is called *hyperbolic* if the Jacobian $\frac{\partial f}{\partial x}(0)$ has no eigenvalues on the imaginary axis. For such equilibria the following theorem holds.

**Theorem 2.5.3** ([Har])  Consider the differential equations (2.5.29) where $f$ denotes a smooth vector field with hyperbolic equilibrium $x = 0$. Then in a neighborhood of $x = 0$ there exist uniquely defined smooth stable and unstable invariant manifolds $S_s$ and $S_u$ passing through $x = 0$ with the same dimensions $n_s$ and $n_u$ as the stable and unstable subspaces $W_s$ and $W_u$ of the linearization (2.5.30), while in $x = 0$, $S_s$ and $S_u$ are tangent to $W_s$ and $W_u$, respectively.

A foliation $\mathcal{F}$ is said to be *invariant* under a complete vector field $f$ if for any leaf $S$ of $\mathcal{F}$ and any $t$ there exists a leaf $\bar{S}_t$ of $\mathcal{F}$ such that

$f^t(S) \subset \tilde{S}_t$ (cf. the definition of invariant distribution). Note that $\tilde{S}_t$ depends on $t$. In fact, it is the manifold obtained by prolongation of $S$ by the vector field $f$ for a certain time $t$. A foliation is said to be *stable* (*unstable*) *invariant* if this foliation is invariant under $f$ and its leaf through $x = 0$ is a stable (unstable) manifold for (2.5.29).

An extension of Theorem 2.5.3 in case the equilibrium of $f$ is not hyperbolic is given in the following theorem.

**Theorem 2.5.4** ([Ca]) Consider again (2.5.29) and assume that $f$ is smooth and $f(0) = 0$. Let $W_s$ and $W_u$ denote the stable and unstable subspaces of (2.5.30) and $W_c$ the subspace associated with the eigenvalues of $\frac{\partial f}{\partial x}(0)$ on the imaginary axis. Then in a neighborhood of $x = 0$ there exist uniquely defined smooth stable and unstable invariant manifolds $S_s$ and $S_u$ and an invariant manifold $S_c$, called the center manifold, passing through $x = 0$ and tangent to $W_s$, $W_u$ and $W_c$, respectively. In general, $S_c$ is not smooth.

The center manifold in Theorem 2.5.4 is not uniquely defined in general. Moreover, the smoothness can be increased by shrinking the region of definition.

We end this section with some remarks on the notion of minimum phase systems. Consider the square system (2.5.1) and assume that $x_0 = 0$ is a regular point of the decoupling matrix. As noted earlier this implies that $\Pi^* = 0$. Suppose that the constrained (or zero) dynamics are given by

$$(2.5.31) \quad \dot{z} = F(z), \quad z \in M_0$$

Let $S_s$ and $S_u$ denote the stable and unstable invariant manifold for (2.5.31) and $S_c$ a center manifold. Let $d_s = \dim S_s$, $d_u = \dim S_u$ and $d_c = \dim S_c$. Then $d := d_s + d_u + d_c$ equals the dimension of the distribution $\Delta^*$. The system (2.5.1) is called *exponentially minimum phase* if $d_u = d_c = 0$. Note that if (2.5.1) is linear, then $d_u = d_c = 0$ implies that $V^*$ and $V_s^*$ coincide, since all transmission zeros are exponentially stable then.

# 3. THE LOCAL DISTURBANCE DECOUPLING PROBLEM WITH STABILITY FOR NONLINEAR SYSTEMS 1

## 3.1 Introduction

In this chapter we consider the Local Disturbance Decoupling Problem with Stability (LDDPS) for nonlinear systems. Our treatment of the problem very much resembles the linear paradigm.

Consider the stabilizable linear system

$$
(3.1.1) \quad
\begin{cases}
\dot{x} = Ax + Bu + Eq, \quad x \in \mathbb{R}^n, \ u \in \mathbb{R}^m, \ q \in \mathbb{R}^r \\
y = Cx, \qquad\qquad y \in \mathbb{R}^\ell
\end{cases}
$$

Recall from Chapter 1 that $V_s^*$ denotes the largest stabilizability subspace in ker $C$ and that the DDPS for (3.1.1) is solvable if and only if

$$(3.1.2) \quad \operatorname{im} E \subset V_s^*$$

Assume that (3.1.2) holds. The DDPS can be solved then by choosing a regular static state feedback

$$(3.1.3) \quad u = Fx + Gv$$

with $F \in \mathcal{F}(V_s^*)$ that also exponentially stabilizes the dynamics modulo $V_s^*$. We will look at this solution in another way now. Assume for simplicity that the system (3.1.1) is controllable. Suppose that the feedback (3.1.3) is chosen in such a way that $F \in \mathcal{F}(V_s^*)$. The linear subspace $V_s^*$ can be considered as an integral manifold through $x = 0$ of the flat distribution $\Delta_{V_s^*}$ (cf. Remark 2.2.6). Now the manifold $V_s^*$ is invariant under the linear vector field $(A+BF)x$, for $x = 0$ is an equilibrium of this vector field. Since $\Delta_{V_s^*}$ is spanned by constant vector fields, this distribution is necessarily invariant under any vector field of the form $X_{ij} := (A+BF)^i(BG)_j$, $i = 0,\ldots,n-1$, $j = 1,\ldots,m$. ($(BG)_j$ denotes the $j$-th column of the matrix $BG$.) This implies that for all $t$ $X_{ij}^t(V_s^*)$ is again an integral manifold of the distribution $\Delta_{V_s^*}$. Of course, $X_{ij}^t(V_s^*) = \bar{x} + V_s^*$ for some $\bar{x} \in \mathbb{R}^n$ (depending on $t$). As a matter of fact, it is possible to construct the foliation $\{\bar{x} + V_s^* \mid \bar{x} \in \mathbb{R}^n\}$ starting from the integral manifold $V_s^*$ through $x = 0$. Indeed, by the controllability of (3.1.1),

(3.1.4)     $sp\{(A+BF)^i(BG)_j \mid i = 0,\ldots,n-1, \; j = 1,\ldots,m\} = T_x\mathbb{R}^n$

for any $x \in \mathbb{R}^n$. Hence, it is possible to find independent vector fields $X_1,\ldots,X_{n-k}$ of the form $(A+BF)^i(BG)_j$ for some $i \in \{0,\ldots,n-1\}$, $j \in \{1,\ldots,m\}$ that are transversal to the $k$-dimensional manifold $V_s^*$. Now the set

(3.1.5)     $\{X_{n-k}^{t_{n-k}} \circ \cdots \circ X_1^{t_1}(V_s^*) \mid t_1,\ldots,t_{n-k} \in \mathbb{R}\}$

defines a foliation on $\mathbb{R}^n$. Note that the order of the $X_i$'s in (3.1.5) does not matter, since $[X_i,X_j] = 0$ for $i,j = 1,\ldots,n-k$. This foliation (3.1.5) coincides with the foliation $\{\bar{x} + V_s^* \mid \bar{x} \in \mathbb{R}^n\}$, because the $X_i$'s are constant vector fields.

We conclude from the preceding that it is possible to construct the foliation (3.1.5) and thus the distribution $\Delta_{V_s^*}$, using only the integral manifold $V_s^*$ through $x = 0$ and an arbitrary set of constant vector fields chosen from (3.1.4) transversal to this manifold.

**Remark 3.1.1** Clearly, the same construction can be followed for any subspace $V$ contained in $V_s^*$ for which $(A+BF)V \subset V$. In particular, the preceding holds for any stabilizability subspace in ker $C$. □

In the next section we introduce stabilizability distributions and we give the construction of a nonlinear analogue $\Delta_s^*$ of $V_s^*$ starting from an integral manifold that is invariant under the (modified) drift vector field. At the end of the section we briefly comment on the differences with the linear case (see Remark 3.2.10). In Section 3.3 the solution of the LDDPS is given using $\Delta_s^*$. Finally, in Section 3.4 it is shown that stabilizability distributions also play a role in the solution of the Strong Local Input-Output Decoupling Problem with Stability.


## 3.2  Stabilizability distributions


Consider the smooth nonlinear system

(3.2.1)     $\begin{cases} \dot{x} = f(x) + g(x)u + p(x)q, \; f(0) = 0, \; x \in \mathbb{R}^n, \; u \in \mathbb{R}^m, \; q \in \mathbb{R}^r \\ y = h(x), \qquad\qquad\qquad\qquad h(0) = 0, \; y \in \mathbb{R}^\ell \end{cases}$

In this section we introduce the concept of stabilizability distribution which plays a key role in the solution of the LDDPS for nonlinear systems. We show that, under certain assumptions, the largest stabilizability distribution in the kernel of the output mapping (denoted by $\Delta_s^*$) exists.

**Definition 3.2.1**  A distribution $\Delta$ is called a *stabilizability distribution* if $\Delta$ is regular, locally controlled invariant and if the linearization of the dynamics $\dot{x} = f(x)+g(x)u$ restricted to the leaf $S_0$ of $\Delta$ through $x = 0$ can be stabilized asymptotically.

**Remark 3.2.2**

(i)  This definition generalizes the concept of stabilizability subspace as introduced in Chapter 1 (see also Lemma 4.3.2). Stabilizability distributions are introduced, because in the LDDPS the requirement for disturbance decoupling restricts the possibilities to stabilize the system. For disturbance decoupling an invariant distribution containing the disturbance vector fields is sought. On the leaf of this distribution $\Delta$ through the equilibrium the dynamics are partly fixed (cf. the linear case in Chapter 1). If $\Delta$ is a stabilizability distribution, then these dynamics are asymptotically stabilizable.

(ii)  Note that a stabilizability distribution is by definition constant dimensional, whereas a locally controlled invariant distribution may be singular. □

Since the definition of a stabilizability distribution is independent of the disturbances $q$ in (3.2.1), we take $q = 0$ in the rest of this section.

For explanatory reasons we first consider square systems that are strongly input–output decouplable. So, assume that $m = \ell$ and that

(A1)  $x = 0$ is a regular point of the decoupling matrix.

It follows from the previous chapter that (A1) implies that the relative degrees are constant and finite, say equal to $r_1, \ldots, r_\ell$, and that the decoupling matrix $A(x)$ has full row rank on a neighborhood $\mathcal{O}$ of $x = 0$. Moreover, $\Pi^* = 0$ for the square system (3.2.1). Without loss of generality, we may assume that the distribution $\Delta^* = \mathrm{sp}\{\frac{\partial}{\partial x_1}\}$ is invariant under $f$ and $g_i$, $i = 1, \ldots, m$, and that the system (3.2.1) has the form

$$(3.2.2) \quad \begin{cases} \dot{x}_1 = \hat{f}^1(x_1,x_2) + \hat{g}^1(x_1,x_2)u \\ \dot{x}_2 = Ax_2 \qquad\qquad + Bu \\ y_i = x^1_{2i}, \quad i = 1,\dots,\ell \end{cases}$$

where

$$(3.2.3) \quad A = \begin{bmatrix} A^1 & & 0 \\ & \ddots & \\ 0 & & A^\ell \end{bmatrix}, \quad B = \begin{bmatrix} B^1 & & 0 \\ & \ddots & \\ 0 & & B^\ell \end{bmatrix}, \quad x_2 = (x^T_{21},\dots,x^T_{2\ell})^T$$

with

$$(3.2.4) \quad A^i = \begin{bmatrix} 0 & 1 & & 0 \\ & \ddots & \ddots & \\ & & & 1 \\ 0 & & & 0 \end{bmatrix}_{r_i \times r_i}, \quad B^i = \begin{bmatrix} 0 \\ \vdots \\ 0 \\ 1 \end{bmatrix}_{r_i}, \quad i = 1,\dots,\ell$$

Obviously, $(A^i,B^i)$ is a controllable pair for $i = 1,\dots,\ell$. Hence it is possible to apply a feedback

$$(3.2.5) \quad u = \begin{bmatrix} N^1 & & 0 \\ & \ddots & \\ 0 & & N^\ell \end{bmatrix} x_2 + v =: Nx_2 + v$$

such that $M := A+BN$ is *anti-stable*, i.e. $\sigma(M) \subset \mathbb{C}^+$, the open right-half of the complex plane. Now system (3.2.2,5) has the form

$$(3.2.6) \quad \begin{cases} \dot{x}_1 = \hat{f}^1(x_1,x_2) + \hat{g}^1(x_1,x_2)Nx_2 + \hat{g}^1(x_1,x_2)v \\ \dot{x}_2 = Mx_2 \qquad\qquad\qquad\qquad + Bv \\ y_i = x^1_{2i}, \quad i = 1,\dots,\ell \end{cases}$$

Since $\Delta^* = \text{sp}\{\frac{\partial}{\partial x_1}\}$, it follows directly from (3.2.6) that

$$(3.2.7) \quad [\tilde{f},\Delta^*] \subset \Delta^*, \quad [\tilde{g}_i,\Delta^*] \subset \Delta^*, \quad i = 1,\dots,m$$

where $\tilde{f}$ and $\tilde{g} = (\tilde{g}_1,\dots,\tilde{g}_m)$ are given by

$$(3.2.8) \quad \tilde{f}(x) = \begin{bmatrix} \hat{f}^1(x_1,x_2) + \hat{g}^1(x_1,x_2)Nx_2 \\ Mx_2 \end{bmatrix}, \quad \tilde{g}(x) = \begin{bmatrix} \hat{g}^1(x_1,x_2) \\ B \end{bmatrix}$$

Recall from Section 2.5 that the restricted zero dynamics are uniquely defined, so $\hat{f}^1(x_1,0)$ does not depend on the "$\alpha$-part" of the feedback (3.2.5). This implies in particular that the set of eigenvalues of the matrix $\dfrac{\partial \hat{f}^1}{\partial x_1}(0,0)$ is fixed. By Theorem 2.5.4 there exist uniquely defined smooth stable and unstable invariant manifolds and an invariant center manifold through $x = 0$ that are invariant under $\tilde{f}$ and tangent to the invariant subspaces $W_s$, $W_u$ and $W_c$ of the matrix

$$(3.2.9) \quad F = \frac{\partial \tilde{f}}{\partial x}(0) = \left[ \begin{array}{cc} \dfrac{\partial \hat{f}^1}{\partial x_1}(x_1,x_2) & \dfrac{\partial \hat{f}^1}{\partial x_2}(x_1,x_2) + \hat{g}^1(x_1,x_2)N \\ 0 & M \end{array} \right]_{x_1=0,\; x_2=0}$$

in $x = 0$. Since, by construction, $\sigma(M) \subset \mathbb{C}^+$, it is obvious that the set of stable eigenvalues of $F$ is contained in the set of eigenvalues of the matrix $\dfrac{\partial \hat{f}^1}{\partial x_1}(0,0)$. Note that the stable invariant manifold $S_0$ is just the set of initial states for which solutions of the system $\dot{x} = \tilde{f}(x)$ tend to zero if $t$ tends to infinity. By the foregoing, $S_0$ is uniquely defined and completely contained in $M_0$, the leaf of $\Delta^*$ through $x = 0$. In the sequel we show that (under an additional assumption) there exists a maximal stabilizability distribution $\Delta_s^*$ in $\Delta^*$ (and thus in ker $dh$) and that the integral manifold of this distribution $\Delta_s^*$ through $x = 0$ is contained in the stable invariant manifold $S_0$.

Let $\mathcal{W}$ denote the set of stabilizability distributions contained in ker $dh$ and invariant under $\tilde{f}$ and $\tilde{g}$ (i.e. invariant under $\tilde{f}$ and $\tilde{g}_i$, $i = 1,\ldots,m$). Since the zero distribution is contained in $\mathcal{W}$, this set is nonempty. Define $\Delta^{\tilde{f},\tilde{g}} := $ inv clos $\{\Sigma\Delta_i \mid \Delta_i \in \mathcal{W}\}$, so $\Delta^{\tilde{f},\tilde{g}}$ denotes the involutive closure of the sum of all stabilizability distributions in ker $dh$ that are invariant under $\tilde{f}$ and $\tilde{g}$. Since all $\Delta_i \in \mathcal{W}$ are contained in $\Delta^*$ and $\Delta^*$ is involutive, $\Delta^{\tilde{f},\tilde{g}}$ is contained in $\Delta^*$. Moreover, all these $\Delta_i$'s are invariant under $\tilde{f}$ and $\tilde{g}$ and by using the Jacobi identity it follows that $\Delta^{\tilde{f},\tilde{g}}$ is invariant under $\tilde{f}$ and $\tilde{g}$. Note that $\Delta^{\tilde{f},\tilde{g}}$ need not be constant dimensional. In order to prove that $\Delta^{\tilde{f},\tilde{g}}$ is a stabilizability distribution we assume that

(A2) $\Delta^{\tilde{f},\tilde{g}}$ has constant dimension on $\mathcal{O}$ and the integral manifold of $\Delta^{\tilde{f},\tilde{g}}$ through $x = 0$ is contained in $S_0$.

Now the following proposition holds.

**Proposition 3.2.3** Consider the smooth square system (3.2.1). Assume that (A1) and (A2) hold. Then there exists a uniquely determined largest stabilizability distribution $\Delta^{\tilde{f},\tilde{g}}$ in ker $dh$ that is invariant under $\tilde{f}$ and $\tilde{g}_i$, $i = 1,\ldots,m$.

**Proof** Obviously, $\Delta^{\tilde{f},\tilde{g}}$ is regular and controlled invariant. Since the leaf of this distribution through $x = 0$ is contained in $S_0$, the dynamics of the system (3.2.1) restricted to this manifold are exponentially stabilizable. $\square$

Note that a priori assumption (A2) holds for this specific $\tilde{f}$ and $\tilde{g}_i$, $i = 1,\ldots,m$. One may wonder if another choice of feedback making $\Delta^*$ invariant could lead to a distribution that does not fulfill (A2). It follows from the following proposition that this is not the case. If (A2) holds for some $\tilde{f}$ and $\tilde{g}$, then it holds for any feedback that makes $\Delta^*$ invariant. This justifies our approach to choose an arbitrary feedback.

**Proposition 3.2.4** The distribution $\Delta^{\tilde{f},\tilde{g}}$ is independent of the smooth regular static state feedback that makes $\Delta^*$ invariant, i.e. if $(\alpha,\beta) \in \mathcal{F}(\Delta^*)$ with $\alpha(0) = 0$, then $(\alpha,\beta) \in \mathcal{F}(\Delta^{\tilde{f},\tilde{g}})$.

**Proof** Suppose that $\Delta^{\tilde{f},\tilde{g}} = sp\{\frac{\partial}{\partial x_{11}}\}$ and $\Delta^* = sp\{\frac{\partial}{\partial x_{11}},\frac{\partial}{\partial x_{12}}\}$ and that both distributions are invariant under $\tilde{f}$ and $\tilde{g}$. Then system (3.2.6) can be rewritten as

$$(3.2.10) \quad \begin{cases} \dot{x}_{11} = \hat{f}^{11}(x_{11},x_{12},x_2) + \hat{g}^{11}(x_{11},x_{12},x_2)Nx_2 + \hat{g}^{11}(x_{11},x_{12},x_2)v \\ \dot{x}_{12} = \hat{f}^{12}(x_{12},x_2) + \hat{g}^{12}(x_{12},x_2)Nx_2 \qquad\qquad + \hat{g}^{12}(x_{12},x_2)v \\ \dot{x}_2 = Mx_2 \qquad\qquad\qquad\qquad\qquad\qquad\qquad\qquad + Bv \\ y_i = x_{2i}^1, \quad i = 1,\ldots,\ell \end{cases}$$

It follows immediately from (3.2.10) that $\Delta^{\tilde{f},\tilde{g}}$ is invariant under any feedback $(\alpha,\beta) \in \mathcal{F}(\Delta^*)$, whether this feedback is linear or nonlinear. This is implied by the fact that every such feedback only depends on $x_2$. $\square$

The next theorem immediately follows from Propositions 3.2.3 and 3.2.4.

**Theorem 3.2.5** Consider the smooth square system (3.2.1). Assume that (A1) and (A2) hold. Then there exists a uniquely defined largest stabiliza-bility distribution contained in the kernel of the output mapping.

In the sequel this distribution is denoted as $\Delta_s^*$. So far, we have a result on the existence of $\Delta_s^*$, but no algorithm to calculate this distribution is available. Consider again (3.2.6) and suppose that the stable invariant manifold $S_0$ has been calculated. Assume that

(A3)  The system (3.2.1) is strongly accessible on $\mathcal{O}$.

Recall from Section 2.5 that (A3) is equivalent to $\mathrm{sp}\{\mathrm{ad}_{\tilde{f}}^k \tilde{g}_i, \ i = 1,\dots,m,$ $k \geq 0\}_{LA}(x) = T_x \mathbb{R}^n$ for all $x \in \mathcal{O}$. We try to construct, starting from $S_0$, a foliation on $\mathcal{O}$ that is invariant under $\tilde{f}$ and $\tilde{g}_i$, $i = 1,\dots,m$. If this is possible the distribution associated with this foliation is equal to $\Delta_s^*$. Let $D$ denote the set of vector fields

(3.2.11)  $D = \{\mathrm{ad}_{\tau_s} \mathrm{ad}_{\tau_{s-1}} \cdots \mathrm{ad}_{\tau_1} \tau_0 | \ s \in \mathbb{N}, \ \tau_0,\dots,\tau_s \in \{\tilde{f},\tilde{g}_1,\dots,\tilde{g}_m\}\}$

If $X$ belongs to $D$ then the manifold $X^t(S_0)$ should be an integral manifold of this distribution $\Delta_s^*$ (for, if $\Delta_s^*$ is invariant under $\tilde{f}$ and $\tilde{g}_i$, $i = 1,\dots,m$, then it is invariant under all Lie brackets of these vector fields). Assume that $S_0$ has dimension $k$. By the accessibility condition (A3) it is possible to find, locally around $x = 0$, independent vector fields $X_1,\dots,X_{n-k}$ in $D$ that are transversal to $S_0$. As a matter of fact, none of these $X_i$'s is equal to $\tilde{f}$, since $\tilde{f}(0) = 0$. Once the order of the $X_i$'s is fixed, the set

(3.2.12)  $\{X_{n-k}^{t_{n-k}} \circ X_{n-k-1}^{t_{n-k-1}} \circ \dots \circ X_1^{t_1}(S_0)| \ -\epsilon \leq t_i \leq \epsilon, \ 1 \leq i \leq n-k\}$

with $\epsilon$ sufficiently small defines a foliation in a neighborhood of $x = 0$. To explain this, we construct the foliation (3.2.12) for a one-dimensional $S_0$ in $\mathbb{R}^3$; the general case follows along the same lines. In this special case there exist locally around $x = 0$ two independent vector fields $X_1$ and $X_2$ in the set $D$ that are transversal to $S_0$. It can easily be seen that the set $\{X_1^{t_1}(S_0)| \ -\epsilon \leq t_1 \leq \epsilon\}$ defines (locally around $x = 0$) a foliation on a two-dimensional manifold $L$ in $\mathbb{R}^3$. Next, consider a point $p$ outside $L$, but sufficiently close to $x = 0$. Then there exists a $t_2$ such that $q := X_2^{-t_2}(p)$ lies on one of the leaves of $\{X_1^{t_1}(S_0) \mid -\epsilon \le t_1 \le \epsilon\}$, say $S_1$. Hence $p \in X_2^{t_2}(S_1) = X_2^{t_2} \circ X_1^{t_1}(S_0)$. Since $p$ is arbitrary, the foliation (3.2.12) is defined on a neighborhood of $x = 0$ in $\mathbb{R}^3$. Note that it depends on the order of the $X_i$'s. If the foliation (3.2.12) is invariant under $\tilde{f}$ and the $\tilde{g}_i$'s, it defines the distribution $\Delta_s^*$ that is invariant under these vector fields. Furthermore, in that case the foliation (3.2.12) does not depend on the order of the $X_i$'s.

If the distribution $\Delta$ that is defined by the foliation (3.2.12) is not invariant under $\tilde{f}$ and $\tilde{g}$, then dim $(\Delta_s^*)$ is strictly less than $k$, the dimension of $S_0$, so $S_0$ is not the leaf of $\Delta_s^*$ through $x = 0$. In that case, it is necessary to search for a lower dimensional manifold $S_1 \subset S_0$ that is invariant under $\tilde{f}$ and to repeat the preceding construction. Unfortunately, there exist many of such manifolds $S_1$ and it is not clear beforehand which one could be taken as a candidate for generating $\Delta_s^*$. Therefore, it seems that in this way $\Delta_s^*$ can be calculated easily only if the construction for $S_0$ works.

**Remark 3.2.6** For the systems considered till now, the term *stabilizability* distribution might be confusing. Maybe, the term *stable* distribution would be more appropriate, because making $\Delta^*$ invariant fixes the restricted zero dynamics (which are equal to the zero dynamics under the assumption (A1)) completely. Hence, "there is nothing left to stabilize".
In case the number of inputs ($m$) is larger than the number of outputs ($\ell$), $\Pi^*$ is not the zero distribution and, as is shown below, the term stabilizability distribution is well chosen. $\qquad\qquad\square$

Consider the system (3.2.1) now with $m > \ell$. In this case, the largest local controllability distribution $\Pi^*$ in ker $dh$ is not the zero distribution. However, the dynamics of the system (3.2.1) restricted to the leaf $L_0$ of $\Pi^*$ through $x = 0$ are not automatically stabilizable (see Example 2.3.23), so in order to be able to stabilize (3.2.1) we need an extra assumption on stabilizability of the dynamics on $L_0$. Assume that (A1) holds and that

(A4) $\Pi^*$ has constant dimension and dim $G = m$ on $\mathcal{O}$.

Without loss of generality, we may assume that $\Delta^*$ and $\Pi^*$ are invariant under $f$ and $g_i$, $i = 1, \ldots, m$ and that

(3.2.13) $\quad \Delta^* \cap G = \mathrm{sp}\{g_1, \ldots, g_{m-\ell}\}$, $\quad \Pi^* = \mathrm{sp}\{\frac{\partial}{\partial x_0}\}$, $\quad \Delta^* = \mathrm{sp}\{\frac{\partial}{\partial x_0}, \frac{\partial}{\partial x_1}\}$

Then the system (3.2.1) has the form

(3.2.14) $\quad \begin{cases} \dot{x}_0 = \hat{f}^0(x_0, x_1, x_2) + \hat{g}^{01}(x_0, x_1, x_2)u^1 + \hat{g}^{02}(x_0, x_1, x_2)u^2 \\ \dot{x}_1 = \hat{f}^1(x_1, x_2) \hspace{4.2cm} + \hat{g}^{12}(x_1, x_2)u^2 \\ \dot{x}_2 = Ax_2 \hspace{5.5cm} + Bu^2 \\ y_i = x^1_{2i}, \quad i = 1, \ldots, \ell \end{cases}$

where $A$ and $B$ are given by (3.2.3) and

(3.2.15) $\quad u^1 = (u_1, \ldots, u_{m-\ell})$, $\quad u^2 = (u_{m-\ell+1}, \ldots, u_m)$

Application of the feedback

(3.2.16) $\quad u^2 = \begin{bmatrix} N^1 & & 0 \\ & \ddots & \\ 0 & & N^\ell \end{bmatrix} x_2 + v^2 =: Nx_2 + v^2$

again makes the dynamics of the system (3.2.1) modulo $\Delta^*$ anti-stable. In order that $\Delta^*_s$ can be defined uniquely, it is necessary that $\Pi^*$ is contained in $\Delta^*_s$ (cf. the linear case in Chapter 1). Therefore, we assume that

(A5) The linearization of the dynamics (3.2.1) restricted to the leaf $L_0$ of $\Pi^*$ through $x = 0$ is stabilizable.

Note that (A5) implies that $(\frac{\partial \hat{f}^0}{\partial x_0}(0,0,0), \hat{g}^{01}(0,0,0))$ is a stabilizable pair. The dynamics of the system (3.2.14,16) restricted to $L_0$ are given by

(3.2.17) $\quad \dot{x}_0 = \hat{f}^0(x_0, 0, 0) + \hat{g}^{01}(x_0, 0, 0)u^1$

Choose a feedback

(3.2.18) $\quad u^1 = \varphi(x_0) + v^1, \quad \varphi(0) = 0$

such that the matrix

(3.2.19) $\quad \frac{\partial}{\partial x_0}(\hat{f}^0(x_0, 0, 0) + \hat{g}^{01}(x_0, 0, 0)\varphi(x_0))_{x=0}$

is asymptotically stable. This implies that the dynamics of the system (3.2.14,16,18) restricted to $L_0$ are locally exponentially stable. The system (3.2.14,16,18) can be rewritten as

(3.2.20) $\quad \dot{x} = \tilde{f}(x) + \tilde{g}^1(x)v^1 + \tilde{g}^2(x)v^2, \quad y = h(x_2)$

with

$$\tilde{f}(x) = \begin{bmatrix} \hat{f}^0(x_0,x_1,x_2) + \hat{g}^{01}(x_0,x_1,x_2)\varphi(x_0) + \hat{g}^{02}(x_0,x_1,x_2)Nx_2 \\ \hat{f}^1(x_1,x_2) + \hat{g}^{12}(x_1,x_2)Nx_2 \\ Mx_2 \end{bmatrix}$$

(3.2.21)

$$\tilde{g}^1(x) = \begin{bmatrix} \hat{g}^{01}(x_0,x_1,x_2) \\ 0 \\ 0 \end{bmatrix}, \quad \tilde{g}^2(x) = \begin{bmatrix} \hat{g}^{02}(x_0,x_1,x_2) \\ \hat{g}^{12}(x_1,x_2) \\ B \end{bmatrix}$$

As before, there exists a uniquely determined smooth stable invariant manifold $S_0$ completely contained in $M_0$, the leaf of $\Delta^*$ through $x = 0$. By construction, $L_0$ is contained in $S_0$. It can be proved now, analogous to the proof of Propositions 3.2.3 and 3.2.4 and Theorem 3.2.5 that there exists a uniquely defined maximal stabilizability distribution $\Delta_s^*$. Moreover, $\Delta_s^*$ necessarily contains $\Pi^*$, since $\Pi^*$ is invariant under $\tilde{f}$ and $\tilde{g}_i$, $i = 1,\ldots,m$ and the dynamics restricted to $L_0$ are locally exponentially stable.

**Theorem 3.2.7** Consider the smooth system (3.2.1). Assume that (A1), (A2), (A4) and (A5) hold. Then there exists a uniquely defined largest stabilizability distribution $\Delta_s^*$ in the kernel of the output mapping. Moreover, $\Pi^*$ is contained in $\Delta_s^*$ and if $\Delta^*$ is invariant under $\tilde{f}$ and $\tilde{g}_i$, $i = 1,\ldots,m$, then so is $\Delta_s^*$.

Up to now, we proved (under certain conditions) the existence of a largest stabilizability distribution in ker $dh$ for strongly input-output decouplable systems only. Hereafter, we extend the results given so far to more general systems.
Observe that in the preceding the dynamics of the system (3.2.1) modulo $\Delta^*$ were made anti-stable in order to find the stable invariant manifold $S_0$ contained in $M_0$. This manifold can also be found in the following way. Consider again the square system (3.2.2). Note that the zero dynamics

manifold $M_0 = \{x | x_2 = 0\}$. Restriction of the dynamics of the system (3.2.2) to $M_0$ yields the (restricted) zero dynamics

(3.2.22) $\quad \dot{x}_1 = \hat{f}^1(x_1, 0)$

By Theorem 2.5.4 there exists a unique stable invariant submanifold $S_0' = \{x_1 \in M_0 | \phi(x_1) = 0\}$ of $M_0$. Then the manifold $\{x | x_2 = 0, \phi(x_1) = 0\}$ is invariant under $\tilde{f}$ (because $\tilde{f}(0) = 0$) and it coincides with the stable invariant manifold $S_0$ found earlier. For the nonsquare system (3.2.14) $S_0$ can be found analogously.

We now turn to systems for which (A1) does not hold.

Consider the smooth system (3.2.1). Assume that (A4) and (A5) hold and that

(A6) $\quad \Delta^*$ and $\Delta^* + G$ are constant dimensional on $\mathcal{O}$.

Once more, we may assume that $\Pi^*$ and $\Delta^*$ are invariant under $f$ and $g_i$, $i = 1, \ldots, m$ and that

(3.2.23) $\quad \Delta^* \cap G = \mathrm{sp}\{g_1, \ldots, g_s\}, \quad \Pi^* = \mathrm{sp}\{\frac{\partial}{\partial x_1}\}, \quad \Delta^* = \mathrm{sp}\{\frac{\partial}{\partial x_1}, \frac{\partial}{\partial x_2}\}$

In that case the system (3.2.1) takes the form

$$
(3.2.24) \quad
\begin{cases}
\dot{x}_1 = \hat{f}^1(x_1, x_2, x_3) + \hat{g}^{11}(x_1, x_2, x_3)u^1 + \hat{g}^{12}(x_1, x_2, x_3)u^2 \\
\dot{x}_2 = \hat{f}^2(x_2, x_3) \qquad\qquad\qquad\qquad\quad + \hat{g}^{22}(x_2, x_3)u^2 \\
\dot{x}_3 = \hat{f}^3(x_3) \qquad\qquad\qquad\qquad\qquad\;\; + \hat{g}^{32}(x_3)u^2 \\
y = h(x_3)
\end{cases}
$$

By (A5) there exists a feedback

(3.2.25) $\quad u^1 = \varphi(x_1) + v^1, \quad \varphi(0) = 0$

such that

(3.2.26) $\quad \frac{\partial}{\partial x_1}\left(\hat{f}^1(x_1, 0, 0) + \hat{g}^{11}(x_1, 0, 0)\varphi(x_1)\right)_{x=0}$

is exponentially stable. Now the dynamics of the system (3.2.24,25) restricted to $M_0 = \{x | x_3 = 0\}$ are given by

$$
(3.2.27) \quad
\begin{cases}
\dot{x}_1 = \hat{f}^1(x_1, x_2, 0) + \hat{g}^{11}(x_1, x_2, 0)\varphi(x_1) + \hat{g}^{11}(x_1, x_2, 0)v^1 \\
\dot{x}_2 = \hat{f}^2(x_2, 0)
\end{cases}
$$

and the restricted zero dynamics (cf. Section 2.5) by

$$(3.2.28) \quad \dot{\bar{x}}_2 = \hat{f}^2(\bar{x}_2, 0)$$

Let $S_0'$ denote the stable invariant manifold of (3.2.28) through the equilibrium $x = 0$, i.e. $S_0' = \{\bar{x}_2 \mid \xi(\bar{x}_2) = 0\}$ for some $\xi$. Then it is clear that the stable invariant manifold $S_0$ of (3.2.24,25) through $x = 0$ is given by $S_0 = \{x \mid x_3 = 0, \xi(x_2) = 0\}$.
The following generalization of Theorem 3.2.7 can easily be proved now.

**Theorem 3.2.8** Consider the smooth square system (3.2.1). Assume that (A2), (A4), (A5) and (A6) hold. Then there exists a uniquely defined largest stabilizability distribution $\Delta_s^*$ contained in the kernel of the output mapping.

**Remark 3.2.9** The approach in this section differs a bit from that in [vdWN2] and [vdW1]. In these articles we assumed that $x = 0$ is a hyperbolic equilibrium of the restricted zero dynamics. Under that condition the integral manifold of $\Delta^{\tilde{f}, \tilde{g}}$ through $x = 0$ is automatically contained in the stable invariant manifold $S_0$. In this section no restriction on the equilibrium is imposed, but then it is necessary to require explicitly that the leaf of $\Delta^{\tilde{f}, \tilde{g}}$ through $x = 0$ is contained in $S_0$. □

**Remark 3.2.10** Although the definitions of stabilizability distributions and stabilizability subspaces are analogous, there is an important difference between $\Delta_s^*$ and $V_s^*$. The dimension of $\Delta_s^*$ may be strictly less than the dimension of the stable invariant manifold $S_0$, whereas in the linear case the dimensions of $S_0$ and $V_s^*$ are always equal. This phenomenon is mainly due to the (strong!) extra requirement that $\Delta_s^*$ should be invariant under the input vector fields $g_i$, $i = 1, \ldots, m$. (Note that in the linear case this requirement is automatically fulfilled by $V_s^*$.) Another extra condition in the nonlinear case is that stabilizability of the linearized dynamics on the leaf of $\Pi^*$ through $x = 0$ is required in order that $\Pi^*$ is contained in $\Delta_s^*$. □

We end up this section by giving three examples. In the first example the dimension of $\Delta_s^*$ is strictly less than the dimension of $S_0$, while in Example 3.2.12 $\Pi^* \neq 0$. In the third example condition (A1) does not hold.

**Example 3.2.11** Consider the system (3.2.1) with $n = 4$, $m = \ell = 1$ and $q = 0$ and

$$(3.2.29) \quad f(x) = \begin{pmatrix} -2x_1 \\ -x_2+x_4 \\ x_3-x_2x_3 \\ 3x_4 \end{pmatrix}, \quad g(x) = \begin{pmatrix} 1 \\ -1 \\ 1 \\ 1 \end{pmatrix}, \quad h(x) = x_4$$

In this case $L_g h(x) = 1$, hence $r = 1$ and $\Delta^* = \mathrm{sp}\{\frac{\partial}{\partial x_1}, \frac{\partial}{\partial x_2}, \frac{\partial}{\partial x_3}\}$. Obviously, $\Delta^*$ is invariant under $f$ and $g$ and

$$(3.2.30) \quad \frac{\partial f}{\partial x}(0) = \begin{pmatrix} -2 & 0 & 0 & 0 \\ 0 & -1 & 0 & 1 \\ 0 & 0 & 1 & 0 \\ 0 & 0 & 0 & 3 \end{pmatrix}$$

and the stable invariant manifold through $x = 0$ is $S_0 = \{x \mid x_3 = x_4 = 0\}$. Define

$$X_1 := [f,g] = \begin{pmatrix} 2 \\ -2 \\ -x_3-1+x_2 \\ -3 \end{pmatrix}, \quad X_2 := [g,X_1] = \begin{pmatrix} 0 \\ 0 \\ -2 \\ 0 \end{pmatrix}$$

(3.2.31)

$$X_3 := [f,X_1] = \begin{pmatrix} 4 \\ 1 \\ -x_2-3x_3+x_4+x_2x_3-(1-x_2)(-x_3-1+x_2) \\ 9 \end{pmatrix}$$

In $x = 0$ we have

$$(3.2.32) \quad \mathrm{rank}[g,X_1,X_2,X_3](0) = 4$$

and so, the system is strongly accessible.
Now starting from $S_0$ we construct a foliation. Let

$$(3.2.33) \quad S_{s,t} := X_2^s \circ g^t(S_0)$$

Since $S_0 = \{(x_{10},x_{20},0,0)^T \mid x_{10} \in \mathbb{R}, \; x_{20} \in \mathbb{R}\}$, $S_{s,t}$ is given by

$$(3.2.34) \quad S_{s,t} := \{(t+x_{10},-t+x_{20},t-2s,t)^T \mid x_{10} \in \mathbb{R}, \; x_{20} \in \mathbb{R}, \; t,s \in \mathbb{R}\}$$

Obviously, the set $\{S_{s,t} \mid s,t \in \mathbb{R}\}$ gives a foliation in $\mathbb{R}^4$. The distribution $\Delta$ associated with it is $\Delta = \mathrm{sp}\{\frac{\partial}{\partial x_1},\frac{\partial}{\partial x_2}\}$. Unfortunately, this $\Delta$ is not invariant under $f$ as follows from

$$(3.2.35)\quad [f,\frac{\partial}{\partial x_2}] = \begin{pmatrix} 0 \\ 1 \\ x_3 \\ 0 \end{pmatrix} \notin \Delta \quad (\text{unless } x_3 = 0)$$

This implies that $\Delta_s^*$ has dimension one at most. Since $\mathrm{sp}\{\frac{\partial}{\partial x_1}\}$ is invariant under $f$ and $g$ and has $S_0^1 = \{x|\ x_2 = x_3 = x_4 = 0\}$ as stable invariant manifold through $x = 0$ it follows that $\Delta_s^* = \mathrm{sp}\{\frac{\partial}{\partial x_1}\}$. $\qquad\square$

**Example 3.2.12** Consider the system (3.2.1) with $n = 4$, $m = 2$, $\ell = 1$, $r = 1$ and

$$(3.2.36)$$
$$f(x) = \begin{bmatrix} -x_1+x_1^2+3x_4 \\ x_2+x_3+2x_4+x_4^2 \\ 2x_3+(1-x_3)x_4 \\ x_4^2 \end{bmatrix}, \quad g(x) = \begin{bmatrix} x_1 & 0 \\ 0 & 1 \\ 0 & 0 \\ 1 & 0 \end{bmatrix}$$

$$p(x) = \begin{bmatrix} x_2 \\ e^{x_3} \\ 0 \\ 0 \end{bmatrix}, \quad h(x) = x_4$$

Since $L_{g_1}h(x) = 1 \neq 0$ for all $x$, we have that $\Delta^* = \mathrm{sp}\{\frac{\partial}{\partial x_1},\frac{\partial}{\partial x_2},\frac{\partial}{\partial x_3}\}$. Furthermore, $A(x) = (L_{g_1}h\ L_{g_2}h)(x) = (1\ 0)$ has full row rank. Note that the system is not state-space linearizable (see [HSuM], [JR]), since the distribution $\mathrm{sp}\{g_1,g_2,\mathrm{ad}_f g_1,\mathrm{ad}_f g_2\} = \mathrm{sp}\{g_1,g_2,\mathrm{ad}_f g_1\}$ is not involutive. Simple, but tedious, calculations show that the system is strongly accessible in a neighborhood of $x = 0$. Note that $\Delta^*$ is already invariant under the vector fields $f$, $g_1$ and $g_2$. From Algorithm 2.3.20 it follows that $\Pi^* = \mathrm{sp}\{\frac{\partial}{\partial x_2}\} = \mathrm{sp}\{g_2\}$. Choose

$$(3.2.37)\quad u_1 = v_1, \quad u_2 = -2x_2+v_2$$

then the system (3.2.1,36,37) has the form

$$(3.2.38)\quad \dot{x} = \tilde{f}(x) + g_1(x)v_1 + g_2(x)v_2 + p(x)q, \quad y = h(x)$$

with

$$(3.2.39) \quad \tilde{f}(x) = \begin{pmatrix} -x_1+x_1^2+3x_4 \\ -x_2+x_3+2x_4+x_4^2 \\ 2x_3+(1-x_3)x_4 \\ x_4^2 \end{pmatrix}$$

Obviously, the stable invariant manifold $S_0$ through $x = 0$ that is invariant under $\tilde{f}$ is given by $S_0 = \{x \mid x_3 = x_4 = 0\}$. The construction given in this section yields $\Delta_s^* = \text{sp}\{\frac{\partial}{\partial x_1}, \frac{\partial}{\partial x_2}\}$.  □

**Example 3.2.13** Consider the system

$$(3.2.40) \quad \begin{cases} \dot{x}_1 = -x_1+2x_3+(1+x_2)u_1, & \dot{x}_2 = x_1-3x_2+(2+x_6)u_1, & \dot{x}_3 = -x_3 \\ \dot{x}_4 = -x_1x_3+x_4+u_1, & \dot{x}_5 = x_2-2x_5+u_2, & \dot{x}_6 = x_6+u_2 \\ y_1 = x_1, & y_2 = x_2 \end{cases}$$

It follows immediately from (3.2.40) that

$$(3.2.41) \quad \dot{y}_1 = \dot{x}_1 = -x_1+2x_3+(1+x_2)u_1, \quad \dot{y}_2 = \dot{x}_2 = x_1-3x_2+(2+x_6)u_1$$

Hence, the relative degrees equal 1. Obviously, the decoupling matrix is given by $A(x) = \begin{pmatrix} 1+x_2 & 0 \\ 2+x_6 & 0 \end{pmatrix}$, so $A(x)$ has rank 1. Application of Algorithm 2.3.9 yields $\Delta^* = \text{sp}\{\frac{\partial}{\partial x_4}, \frac{\partial}{\partial x_5}\}$. It is clear from (3.2.40) that $\Delta^* \cap G = 0$, so $\Pi^* = 0$. The dynamics of the system (3.2.40) restricted to the leaf of $\Delta^*$ through $x = 0$ are given by

$$(3.2.42) \quad \dot{x}_4 = x_4, \quad \dot{x}_5 = -2x_5$$

This implies that $S_0 = \{x \mid x_i = 0, \, i = 1,2,3,4,6\}$ and $\Delta_s^* = \text{sp}\{\frac{\partial}{\partial x_5}\}$. It can easily be checked that the feedback

$$(3.2.43) \quad u_1 = -2x_4+v_1, \quad u_2 = -2x_6+v_2$$

solves the LDDPS for (3.2.40).  □

## 3.3 The Local Disturbance Decoupling Problem with Stability

Consider again the system (3.2.1) (including the disturbances). Having introduced the concept of stabilizability distribution in the previous section, we show in this section that the solution of the LDDPS for the system (3.2.1) is completely analogous to the solution for linear systems.

**Theorem 3.3.1** Consider the smooth system (3.2.1). Assume that (A2), (A4), (A5) and (A6) hold and that

(A7) The linearization of the system dynamics (3.2.1) modulo $\Delta_s^*$ is stabilizable.

Then the Local Disturbance Decoupling Problem with Stability is solvable if $P \subset \Delta_s^*$. On the other hand, if the LDDPS for (3.2.1) is solvable by making a regular distribution $\Delta$ invariant, then $P \subset \Delta_s^*$.

**Proof**

(i) Assume that $P \subset \Delta_s^*$. Choose $u = \alpha(x) + \beta(x)v$ with $\alpha(0) = 0$ and $\beta(x)$ invertible in a neighborhood $\mathcal{O}$ of $x = 0$ such that $\Delta^*$ and $\Delta_s^*$ are invariant under $\tilde{f} := f + g\alpha$ and $\tilde{g}_i := (g\beta)_i$, $i = 1, \ldots, m$, and such that the drift dynamics restricted to the leaf $L_0$ of $\Pi^*$ through $x = 0$ are exponentially stable. In suitable coordinates the system with feedback takes the form

$$
(3.3.1) \quad
\begin{cases}
\dot{z}_1 = \tilde{f}^1(z_1,z_2,z_3) + \tilde{g}^{11}(z_1,z_2,z_3)v^1 + \tilde{g}^{12}(z_1,z_2,z_3)v^2 + p^1(z_1,z_2,z_3)q \\
\dot{z}_2 = \tilde{f}^2(z_2,z_3) \qquad\qquad\quad\, + \tilde{g}^{22}(z_2,z_3)v^2 \quad\; + p^2(z_1,z_2,z_3)q \\
\dot{z}_3 = \tilde{f}^3(z_3) \qquad\qquad\qquad\, + \tilde{g}^{32}(z_3)v^2 \quad\;\; + p^3(z_1,z_2,z_3)q \\
y = h(z_3)
\end{cases}
$$

with $\Delta_s^* = \mathrm{sp}\{\frac{\partial}{\partial z_1}\}$ and $\Delta^* = \mathrm{sp}\{\frac{\partial}{\partial z_1},\frac{\partial}{\partial z_2}\}$. Since $P \subset \Delta_s^*$, it is clear that $p^2$ and $p^3$ are identically equal to zero. Now $\sigma(\frac{\partial \tilde{f}^1}{\partial z_1}(0,0,0)) \subset \mathbb{C}^-$ and, by (A7),

$$
(3.3.2) \quad (F,G) := \left( \begin{bmatrix} \dfrac{\partial \tilde{f}^2}{\partial z_2}(0,0) & \dfrac{\partial \tilde{f}^2}{\partial z_3}(0,0) \\[2ex] 0 & \dfrac{\partial \tilde{f}^3}{\partial z_3}(0) \end{bmatrix}, \begin{bmatrix} \tilde{g}^{22}(0,0) \\[2ex] \tilde{g}^{32}(0) \end{bmatrix} \right)
$$

is a stabilizable pair. Choose a feedback

$$
(3.3.3) \quad v^2 = Kz_2 + Lz_3 + \tilde{v}^2 = (K\; L)\begin{bmatrix} z_2 \\ z_3 \end{bmatrix} + \tilde{v}^2
$$

such that the matrix $F+G(K\; L)$ is asymptotically stable. This feedback solves the LDDPS for (3.3.1) ((3.2.1)), because it leaves $\Delta_s^*$ invariant.

(ii) Assume that the LDDPS for (3.2.1) is solvable by making the regular distribution $\Delta$ invariant. Then $P \subset \Delta \subset \ker dh$. Moreover, the linearization of the dynamics (3.2.1) restricted to the leaf $S_0$ of $\Delta$ through $x = 0$ can be stabilized exponentially. Hence, $\Delta$ is a stabilizability distribution, so $P \subset \Delta \subset \Delta_s^*$ □

**Remark 3.3.2** The condition (A7) is immediately implied by the assumption that the linearization of the system (3.2.1) is stabilizable and the fact that $\Delta_s^*$ is a regular locally controlled invariant distribution. For details, See Section 4.3 □

**Remark 3.3.3** For the stabilizable linear system influenced by nonlinear disturbances

$$(3.3.4) \quad \begin{cases} \dot{x} = Ax + Bu + p(x)q, \ x \in \mathbb{R}^n, \ u \in \mathbb{R}^m, \ q \in \mathbb{R}^r \\ y = Cx, \qquad\qquad y \in \mathbb{R}^\ell \end{cases}$$

we have that $\Delta_s^*$ and $\Delta_{\mathcal{V}_s^*}$ coincide. (As a matter of fact, none of the assumptions (A1) up to (A7) is needed to prove this (see also Remark 3.2.10).) Theorem 3.3.1 states that the LDDPS for (3.3.4) is solvable if $P \subset \Delta_{\mathcal{V}_s^*}$ and that if the LDDPS is solvable by making a constant dimensional distribution invariant then $P \subset \Delta_{\mathcal{V}_s^*}$. In both cases the LDDPS is solvable by applying a *linear* feedback. This is in agreement with Theorem 2.2 in [vdWN1]. □

We end up with an example.

**Example 3.3.4** (cf. Example 3.2.12) Consider again (3.2.1,36,37). Since $\Delta_s^* = \mathrm{sp}\{\frac{\partial}{\partial x_1}, \frac{\partial}{\partial x_2}\}$ and the linearization of (3.2.38) modulo $\Delta_s^*$ is given by

$$(3.3.5) \quad \dot{z}_1 = 2z_1 + z_2, \quad \dot{z}_2 = w_1$$

it is obvious that e.g. the feedback

$$(3.3.6) \quad v_1 = -9x_3 - 4x_4 + \tilde{v}_1, \quad v_2 = \tilde{v}_2$$

solves the LDDPS for the system (3.2.38). □

## 3.4 The Strong Local Input-Output Decoupling Problem with Stability

In this section we show that stabilizability distributions also play a role in the solution of the problem of achieving strong input-output decoupling and stability at the same time.

Consider the square smooth system

$$(3.4.1) \quad \begin{cases} \dot{x} = f(x) + g(x)u, \ f(0) = 0, \ x \in \mathbb{R}^n, \ u \in \mathbb{R}^m \\ y = h(x), \qquad\qquad h(0) = 0, \ y \in \mathbb{R}^m \end{cases}$$

The problem we consider here is defined as follows (cf. Definition 2.3.15):

**Definition 3.4.1** *Strong Local Input-Output Decoupling Problem with Stability* (SLIODPS) Consider the square smooth system (3.4.1). Under what conditions can we find a smooth regular static state feedback

$$(3.4.2) \quad u = \alpha(x) + \beta(x)v \ , \quad u \in \mathbb{R}^m, \ v \in \mathbb{R}^m, \ x \in \mathbb{R}^n$$

with $\alpha(0) = 0$ defined on a neighborhood $\mathcal{O}$ of $x = 0$ such that the feedback system (3.4.1,2) is strongly input-output decoupled on $\mathcal{O}$ and such that $x = 0$ is a locally exponentially stable equilibrium of the drift dynamics $\dot{x} = f(x)+g(x)\alpha(x)$?

Let $\hat{\Delta}_i^*$ and $\Delta_i^*$ ($i = 1,\ldots,m$) denote the largest local controllability distributions contained in $\underset{j \neq i}{\cap} \ker dh_j$ and $\ker dh_i$, respectively. Suppose that the $\hat{\Delta}_i^*$'s are constant dimensional on a neighborhood $\mathcal{O}$ of $x = 0$. Assume, moreover, that $\tilde{\Delta} := \overset{m}{\underset{i=1}{\cap}} \Delta_i^*$ is a regular distribution and that the noninteraction condition holds on $\mathcal{O}$, i.e.

$$(A8) \quad G = G \cap \hat{\Delta}_1^* + \cdots + G \cap \hat{\Delta}_m^*$$

It is proved in [NS] that (under some regularity assumptions on the $\hat{\Delta}_i^*$'s) in case (A3) holds and dim $G = m$ and dim $dh = m$, then the SLIODP (without stability) is solvable if and only if (A8) holds. It is shown in [DM] that (A8) is equivalent to (A1). As for the SLIODPS it is shown in [IG] (and [GI]) that a necessary condition for stabilizability is that the dynamics of the system (3.4.1) restricted to the leaf $N_0$ of $\tilde{\Delta}$ through $x = 0$ are asymptotically stable.

**Remark 3.4.2** Under the condition (A8) the locally controlled invariant distribution $\tilde{\Delta}$ is equal to the *radical* $\Delta^{r}$ ([Wo]) defined by

$$(3.4.3) \qquad \Delta^{r} := \sum_{i=1}^{m} [\hat{\Delta}_{i}^{*} \cap \sum_{j \neq i} \hat{\Delta}_{j}^{*}] \qquad\qquad \square$$

Let the induced dynamics on $N_0$ be given by

$$(3.4.4) \qquad \dot{z} = \tilde{F}(z), \qquad z \in N_0$$

It follows from [IG] that the dynamics (3.4.4) are independent of the regular static state feedback (3.4.2) with $\alpha(0) = 0$ that decouples the system. Suppose that $z = 0$ is a locally asymptotically stable equilibrium of the dynamics (3.4.4). Generically, the asymptotic stability is not critical, but exponential. In this generic case, $N_0$ is a stable invariant manifold. Since $N_0$ is a leaf of the locally controlled invariant constant dimensional distribution $\tilde{\Delta}$, it follows that $\tilde{\Delta}$ is a stabilizability distribution. (Note that prolongation of $N_0$ as in Section 3.2 would succeed.) This proves part of the following proposition.

**Proposition 3.4.3** Consider the square smooth system (3.4.1). Assume that (A1) (or (A8)) holds and that the linearization of the dynamics (3.4.1) modulo $\tilde{\Delta}$ is stabilizable. Then the Strong Local Input–Output Decoupling Problem with Stability is solvable if and only if $\tilde{\Delta}$ is a stabilizability distribution.

This proposition can be proved by a simple adaptation of the proof of Theorem 3.12 in [GI]. However, note that we require that the stability of the equilibrium is exponential, while Grizzle and Isidori allow for critical stability. Note that Proposition 3.4.3 generalizes the well-known result for controllable linear systems that if the system is input–output decouplable and the dynamics restricted to the radical are asymptotically stable, then the input–output decoupling problem with stability is solvable (see [Wo]).

4. THE LOCAL DISTURBANCE DECOUPLING PROBLEM
   WITH STABILITY FOR NONLINEAR SYSTEMS 2

## 4.1 Introduction

In the preceding chapter we introduced the concept of stabilizability
distribution as a tool to solve the LDDPS. Unfortunately, it appeared that
in general the largest stabilizability distribution $\Delta_s^*$ in the kernel of the
output mapping is hard to calculate analytically. Therefore, we introduce
another method to solve the LDDPS in the present chapter. This more
"problem oriented" method is again motivated by the linear paradigm. To
make this plausible, consider again the stabilizable linear system

$$(4.1.1) \quad \begin{cases} \dot{x} = Ax + Bu + Eq, \quad x \in \mathbb{R}^n, \ u \in \mathbb{R}^m, \ q \in \mathbb{R}^r \\ y = Cx, \qquad\qquad y \in \mathbb{R}^\ell \end{cases}$$

and assume that the DDP for (4.1.1) is solvable, so

$$(4.1.2) \quad \text{im } E \subset V^*$$

Also for linear systems the calculation of the largest stabilizability
subspace $V_s^*$ in ker $C$ is not simple, because it requires the calculation of
eigenspaces which is numerically inattractive. To circumvent this problem,
Basile, Marro and Piazzi ([BMP]) give an alternative procedure to solve the
DDPS for (4.1.1). Before we give their solution, we need some notation. Let
$V_{B,E}^* := V^*(A, \text{im } B + \text{im } E, \text{ker } C)$ and $\mathcal{R}_{B,E}^* := \mathcal{R}^*(A, \text{im } B + \text{im } E, \text{ker } C)$ denote
the largest controlled invariant subspace and the largest controllability
subspace in ker $C$, respectively, in case both $u$ and $q$ are considered as
control inputs. (Note that in this notation $V^* - V_B^*$ and $\mathcal{R}^* - \mathcal{R}_B^*$.)
Analogously, let $\mathcal{S}_*^{B,E} := \mathcal{S}_*(A, \text{im } B + \text{im } E, \text{ker } C)$ denote the smallest
conditioned invariant subspace containing im $B$ + im $E$ (see Definition
1.3(b)).
In [BMP] it is proved that the DDPS for (4.1.1) is solvable if and only if
the dynamics of (4.1.1) restricted to $\hat{V}_E := V^* \cap \mathcal{S}_*^{B,E}$ can be stabilized
asymptotically. In other words, the DDPS for (4.1.1) is solvable if and
only if $\hat{V}_E$ is contained in $V_s^*$ (and hence is a stabilizability subspace).
Note that the subspace $\hat{V}_E$ explicitly depends on the disturbance vectors in

im $E$ contrary to $V_s^*$ that is defined independently of im $E$. Before generalizing this idea to the nonlinear context, we show that $\hat{V}_E$ can be calculated more easily. In fact, we will show that $\hat{V}_E$ equals the smallest controlled invariant subspace in ker $C$ containing $\mathcal{R}^*$ and im $E$ (see Theorem 4.1.5). Note that in general the smallest controlled invariant subspace in ker $C$ containing im $E$ does not exist. In order that this subspace is well-defined, it is crucial that it contains $\mathcal{R}^*$. Let $F \in \mathcal{F}(V^*)$, so $(A+BF)V^* \subset V^*$. Define

(4.1.3)  $\tilde{V}_E := \mathcal{R}^* + \text{im } E + (A+BF)\text{im } E + \ldots = \mathcal{R}^* + < A+BF | \text{ im } E >$

(It can easily be shown that $\tilde{V}_E$ is independent of $F$.) Since (4.1.2) holds, $\tilde{V}_E$ is controlled invariant and contained in ker $C$, so $\tilde{V}_E \subset V^*$.

The subspace $\tilde{V}_E$ as defined in (4.1.3) can be calculated by means of the following algorithm.

**Algorithm 4.1.1**

(4.1.4)  $V_0 := \text{im } E + \mathcal{R}^*$,  $V_{k+1} := V_k + (A+BF)V_k$,  $\tilde{V}_E := V_n$

It follows from (4.1.2) and the fact that $A+BF$ restricted to $V^*/\mathcal{R}^*$ is unique that $\tilde{V}_E$ is the smallest controlled invariant subspace in ker $C$ that contains im $E$ and $\mathcal{R}^*$. The following lemmas are needed to prove equality of $\hat{V}_E$ and $\tilde{V}_E$.

**Lemma 4.1.2**  Assume that (4.1.2) holds. Then $V_{B,E}^* = V^*$.

**Proof**  It follows immediately from $AV^* \subset V^* + \text{im } B$, the maximality of $V^*$ and $V_{B,E}^*$ and

(4.1.5)  $AV_{B,E}^* \subset V_{B,E}^* + \text{im } B + \text{im } E$

that $V^* \subset V_{B,E}^*$. Moreover, by (4.1.2), it follows from (4.1.5) that $AV_{B,E}^* \subset V_{B,E}^* + \text{im } B$. Hence $V_{B,E}^* \subset V^*$. This proves the claim.  □

**Lemma 4.1.3**  Suppose that (4.1.2) holds. Then $\hat{V}_E = \mathcal{R}_{B,E}^*$.

**Proof**  By Lemma 4.1.2 we have that

(4.1.6)    $\hat{V}_E = V^* \cap \mathcal{S}_*^{B,E} = V_{B,E}^* \cap \mathcal{S}_*^{B,E} = \mathcal{R}_{B,E}^*$

The last equality follows from [BM2].                                    □

Note that $\mathcal{R}_{B,E}^*$ can be calculated by the following slightly different version of Algorithm 1.10 (cf. [Wo]).

**Algorithm 4.1.4**

(4.1.7)
$$
\begin{cases}
\mathcal{R}_0 \;\; := (\text{im } B + \text{im } E) \cap V^* \\[2mm]
\mathcal{R}_{k+1} := ((A+BF)\mathcal{R}_k + \text{im } B + \text{im } E) \cap V^* \\[2mm]
\mathcal{R}_{B,E}^* := \mathcal{R}_n
\end{cases}
$$

**Theorem 4.1.5**  Consider the system (4.1.1). Assume that (4.1.2) holds. Then $\hat{V}_E$ equals $\tilde{V}_E$.

**Proof**  It follows from Lemma 4.1.3 that it is sufficient to show that $\tilde{V}_E$ equals $\mathcal{R}_{B,E}^*$. This can easily be established by using the Algorithms 4.1.1 and 4.1.4. By definition, $\mathcal{R}_{B,E}^*$ is controlled invariant, contained in ker $C$ and contains both im $E$ and $\mathcal{R}^*$, since im $B \subset$ im $B +$ im $E$. Hence, by definition of $\tilde{V}_E$, $\tilde{V}_E \subset \mathcal{R}_{B,E}^*$. On the other hand

(4.1.8)    $\mathcal{R}_0 = (\text{im } B + \text{im } E) \cap V^* = \text{im } B \cap V^* + \text{im } E \subset \mathcal{R}^* + \text{im } E = V_0$

Suppose that $\mathcal{R}_i \subset V_i$ for all $i \le k$. Then

(4.1.9) $\mathcal{R}_{k+1} = V^* \cap ((A+BF)\mathcal{R}_k + \text{im } B + \text{im } E) \subset V^* \cap ((A+BF)V_k + \text{im } B + \text{im } E)$

$= V^* \cap \text{im } B + ((A+BF)V_k + \text{im } E) \subset \mathcal{R}^* + \text{im } E + (A+BF)V_k$

$= V_0 + (A+BF)V_k \subset V_{k+1}$

This immediately implies that $\mathcal{R}_{B,E}^* \subset \tilde{V}_E$. We conclude that $\tilde{V}_E$ equals $\hat{V}_E$.   □

In the next section we show, motivated by this solution of the DDPS, that the distribution $(\Delta^P)_*$, a nonlinear analogue of $\tilde{V}_E$, plays a similar role in the solution of the LDDPS for nonlinear systems. In Section 4.3 we compare the assumptions under which the LDDPS is solved in Chapter 3 and in Section 4.2 to each other. In Section 4.4 a related result in the literature is discussed.

## 4.2  The Local Disturbance Decoupling Problem with Stability

Consider again the smooth nonlinear system

$$(4.2.1) \quad \begin{cases} \dot{x} = f(x) + g(x)u + p(x)q, & f(0) = 0, \ x \in \mathbb{R}^n, \ u \in \mathbb{R}^m, \ q \in \mathbb{R}^r \\ y = h(x), & h(0) = 0, \ y \in \mathbb{R}^\ell \end{cases}$$

Assume that

(B1)  $P \subset \Delta^*$ on a neighborhood $\mathcal{O}$ of $x = 0$;

(B2)  dim $G = m$ and $\Delta^*$ is constant dimensional on $\mathcal{O}$.

Choose a regular static state feedback

$$(4.2.2) \quad u = \alpha(x) + \beta(x)v, \quad \alpha(0) = 0, \quad \beta(x) \text{ invertible on } \mathcal{O}$$

such that for the feedback modified system (4.2.1,2) we have that $\Delta^*$ is invariant under $\tilde{f} := f + g\alpha$ and $\tilde{g}_i := (g\beta)_i$, $i = 1, \ldots, m$, so $(\alpha, \beta) \in \mathcal{F}(\Delta^*)$. Let $\Pi^*$ again denote the largest local controllability distribution in ker $dh$. We define a distribution $\Delta_{\tilde{f}, \tilde{g}}^P$ by the following algorithm.

**Algorithm 4.2.1**

$$(4.2.3) \quad \begin{cases} \Delta_0 := P + \Pi^* \\ \Delta_{k+1} := \Delta_k + [\tilde{f}, \Delta_k] + \sum_{i=1}^{m} [\tilde{g}_i, \Delta_k], \quad k = 1, 2, \ldots \end{cases}$$

Let $\Delta'$ denote the sum of all distributions $\Delta_k$, $k = 0, 1, \ldots$ . (Note that if all distributions $\Delta_k$ are constant dimensional, then the algorithm converges in at most $n$ steps.) Let $\Delta_{\tilde{f}, \tilde{g}}^P$ denote the involutive closure of $\Delta'$. Assume that

(B3)  $\Delta_{\tilde{f}, \tilde{g}}^P$ is constant dimensional on $\mathcal{O}$.

Obviously, $\Delta_{\tilde{f}, \tilde{g}}^P$ is contained in $\Delta^*$ and locally controlled invariant. (Recall from Definition 2.3.4 that a controlled invariant distribution is always involutive.) A priori the distribution $\Delta_{\tilde{f}, \tilde{g}}^P$ depends on the choice of the feedback (4.2.2). It follows from the following theorem that in fact it does not depend on this feedback (cf. similar comments w.r.t. the distribution $\Delta^{\widetilde{f},\widetilde{g}}$ in Section 3.2).


**Theorem 4.2.2** Consider the system (4.2.1). Assume that (B1), (B2) and (B3) hold. Then $\Delta^{P}_{\widetilde{f},\widetilde{g}}$ is independent of the choice of the feedback (4.2.2) with $(\alpha,\beta) \in \mathcal{F}(\Delta^{*})$. Moreover, $\Delta^{P}_{\widetilde{f},\widetilde{g}}$ is the smallest constant dimensional locally controlled invariant distribution in the kernel of the output mapping that contains $\Pi^{*}$ and the disturbance vector fields $P$.


**Proof** Since $\Delta^{P}_{\widetilde{f},\widetilde{g}}$ and $\Delta^{*}$ are involutive and constant dimensional, there exists a coordinate transformation $z = \varphi(x)$ on $\mathcal{O}$ such that $\Delta^{P}_{\widetilde{f},\widetilde{g}} = \mathrm{sp}\{\frac{\partial}{\partial z_1}\}$, $\Delta^{*} = \mathrm{sp}\{\frac{\partial}{\partial z_1},\frac{\partial}{\partial z_2}\}$. In these coordinates the system (4.2.1,2) has the following form:

(4.2.4)
$$
\begin{cases}
\dot{z}_1 = \widetilde{f}^1(z_1,z_2,z_3) + \widetilde{g}^{11}(z_1,z_2,z_3)v^1 + \widetilde{g}^{12}(z_1,z_2,z_3)v^2 \\
\qquad\qquad\qquad\qquad\qquad\qquad\qquad\qquad\qquad\quad + p^1(z_1,z_2,z_3)q \\
\dot{z}_2 = \widetilde{f}^2(z_2,z_3) \qquad\qquad\qquad\qquad\quad + \widetilde{g}^{22}(z_2,z_3)v^2 \\
\dot{z}_3 = \widetilde{f}^3(z_3) \qquad\qquad\qquad\qquad\qquad\quad + \widetilde{g}^{32}(z_3)v^2 \\
y = h(z_3)
\end{cases}
$$

where $G \cap \Delta^{*} = \mathrm{sp}\{\widetilde{g}^1\} = \mathrm{sp}\{\widetilde{g}_1,\ldots,\widetilde{g}_s\}$ and $\Delta^{*} \cap \mathrm{sp}\{\widetilde{g}^2\} = 0$. Since dim $G = m$, we have that dim $\mathrm{sp}\{\widetilde{g}^{32}_{s+1},\ldots,\widetilde{g}^{32}_m\} = m - s$. Hence, any additional locally defined regular static state feedback $v = \widetilde{\alpha}(z) + \widetilde{\beta}(z)w$ with $\widetilde{\alpha}(0) = 0$ and $(\widetilde{\alpha},\widetilde{\beta}) \in \mathcal{F}(\Delta^{*})$ fulfills the condition $v^2 = \widetilde{\alpha}_2(z_3) + \widetilde{\beta}_{22}(z_3)w^2$. This implies that application of such a feedback to (4.2.4) does not influence the structure of the equations (4.2.4). In other words, the distribution $\Delta^{P}_{\widetilde{f},\widetilde{g}}$ is invariant under $\widetilde{f} + \widetilde{g}\widetilde{\alpha}$ and $\widetilde{g}\widetilde{\beta}$ and contains $P$ and $\Pi^{*}$. Hence if the smallest involutive distribution with these properties $(\Delta^{P}_{\widetilde{f}+\widetilde{g}\widetilde{\alpha},\widetilde{g}\widetilde{\beta}})$ is constant dimensional, then it is contained in $\Delta^{P}_{\widetilde{f},\widetilde{g}}$. By converting arguments it follows that $\Delta^{P}_{\widetilde{f},\widetilde{g}}$ equals $\Delta^{P}_{\widetilde{f}+\widetilde{g}\widetilde{\alpha},\widetilde{g}\widetilde{\beta}}$. So the distribution $\Delta^{P}_{\widetilde{f},\widetilde{g}}$ is independent of the feedback (4.2.2) as long as (4.2.2) is such that $\Delta^{*}$ is invariant under $\widetilde{f}$ and $\widetilde{g}$. Hence $\Delta^{P}_{\widetilde{f},\widetilde{g}}$ equals $\Delta^{P}_{\widetilde{f},\widetilde{g}}$ for every $\widetilde{f}$ and $\widetilde{g}$ that leave $\Delta^{*}$ invariant. Now, suppose that $\Delta$ is a constant dimensional locally controlled invariant distribution in ker $dh$ containing $P$ and $\Pi^{*}$. Assume that $\Delta$ is invariant under $\bar{f}$ and $\bar{g}$, but $\Delta^{*}$ is *not*. Since $\Delta \subset \Delta^{*}$ and both distributions are constant dimensional and involutive there exists a coordinate transformation $z = \varphi(x)$ such that $\Delta = \mathrm{sp}\{\frac{\partial}{\partial z_1}\}$, $\Delta^{*} = \mathrm{sp}\{\frac{\partial}{\partial z_1},\frac{\partial}{\partial z_2}\}$. Moreover, there exists a feedback $v = \bar{\alpha}(z) + \bar{\beta}(z)w$ with $\bar{\alpha}(0) = 0$, $\bar{\beta}(z)$

62

invertible on $\mathcal{O}$, such that $\Delta$ and $\Delta^*$ are invariant under $\bar{\bar{f}}:= \bar{f} + \bar{g}\alpha$ and $\bar{\bar{g}}:= \bar{g}\bar{\beta}$ and $\frac{\partial\bar{\bar{f}}}{\partial z_1} - \frac{\partial\bar{f}}{\partial z_1}$ and $\frac{\partial\bar{\bar{g}}}{\partial z_1} - \frac{\partial\bar{g}}{\partial z_1}$ (see [Nij2]). Since $\Delta$ is invariant under $\bar{f}$ and $\bar{g}$ and contains $P$ and $\Pi^*$, it is obvious that $\Delta^P_{\bar{f},\bar{g}} = \Delta^P_{\bar{\bar{f}},\bar{\bar{g}}} \sim$ (which is constant dimensional by (B3)) is contained in $\Delta$. □

**Corollary 4.2.3** Consider the system (4.2.1). Assume that (B1), (B2) and (B3) hold. Then there exists a uniquely defined smallest constant dimensional locally controlled invariant distribution $(\Delta^P)_*$ in ker $dh$ that contains the disturbance vector fields $P$ and the largest local controllability distribution $\Pi^*$ in ker $dh$.

The solution of the LDDPS in terms of $(\Delta^P)_*$ is now straightforward.

**Theorem 4.2.4** Consider the smooth system (4.2.1). Assume that (B1), (B2) and (B3) hold and that

(B4)  The dynamics (4.2.1) restricted to the leaf of $(\Delta^P)_*$ through $x = 0$ can be stabilized exponentially;

(B5)  The linearization of the system dynamics (4.2.1) modulo $(\Delta^P)_*$ is stabilizable.

Then the Local Disturbance Decoupling Problem with Stability for (4.2.1) is solvable. On the other hand, if the LDDPS for (4.2.1) is solvable by making a regular distribution $\Delta$ invariant, then the dynamics of the system restricted to the leaf of $\Delta$ through $x = 0$ can be stabilized exponentially and the linearization of the dynamics modulo $\Delta$ is stabilizable.

The simple proof is omitted.

**Remark 4.2.5**
(i) In case $\Pi^*$ equals the zero distribution, the drift dynamics on $S_0$ are fixed, because the restricted zero dynamics are (see also [IG]). In order that the LDDPS is solvable, these dynamics have to be exponentially stable. In case $\Pi^* \neq 0$ the dynamics restricted to $L_0$, the leaf of $\Pi^*$ through $x = 0$ can be stabilized exponentially if the linearization of the system restricted to $L_0$ is stabilizable (cf. Section 3.2).
(ii) The distributions $\Delta$ and $(\Delta^P)_*$ in Theorem 4.2.4 are stabilizability distributions (see Definition 3.2.1). □

The following example illustrates the theory developed so far.

**Example 4.2.6** (cf. Example 3.2.12). Consider again the system (4.2.1) with
$n = 4$, $m = 2$, $\ell = r = 1$ and

$$
(4.2.5) \quad f(x) = \begin{bmatrix} -x_1+x_1^2+3x_4 \\ x_2+x_3+2x_4+x_4^2 \\ 2x_3+(1-x_3)x_4 \\ x_4^2 \end{bmatrix}, \quad g(x) = \begin{bmatrix} x_1 & 0 \\ 0 & 1 \\ 0 & 0 \\ 1 & 0 \end{bmatrix}
$$

$$
p(x) = \begin{bmatrix} x_2 \\ e^{x_3} \\ 0 \\ 0 \end{bmatrix}, \quad h(x) = x_4
$$

It can easily be checked that application of Algorithm 4.2.1 to this system
yields $\Delta_0 = \mathrm{sp}\{\frac{\partial}{\partial x_2}, x_2\frac{\partial}{\partial x_1} + e^{x_3}\frac{\partial}{\partial x_2}\}$ and $\Delta_1 = \mathrm{sp}\{\frac{\partial}{\partial x_1}, \frac{\partial}{\partial x_2}\} = (\Delta^P)_*$. The LDDPS
can be solved now by choosing the feedback (3.2.37,3.6). Note that $\Delta_s^*$
equals $(\Delta^P)_*$. $\quad\square$

**Remark 4.2.7** In Example 3.2.12 $(\Delta^P)_*$ and $\Delta_s^*$ coincide. This is not
necessarily the case in general. Choose e.g. $p(x) = \frac{\partial}{\partial x_2}$ in (4.2.5), then
$(\Delta^P)_* = \mathrm{sp}\{\frac{\partial}{\partial x_2}\}$ $(= \Pi^*)$. $\quad\square$

Until now we considered systems of the form (4.2.1) assuming that the
disturbances cannot be measured in contrast to the states. This implied
that we could not allow the feedbacks to depend on $q$. In the rest of this
section we consider the disturbance decoupling problem (with stability) in
case the disturbances are measurable. Before stating the problem
formulation we give a definition. A feedback

$$(4.2.6) \quad u = \alpha(x) + \beta(x)v + \gamma(x)q, \quad u \in \mathbb{R}^m, v \in \mathbb{R}^m, x \in \mathbb{R}^n, q \in \mathbb{R}^r$$

with $\alpha: \mathbb{R}^n \to \mathbb{R}^m$, $\beta: \mathbb{R}^n \to \mathbb{R}^{m \times m}$, $\gamma: \mathbb{R}^n \to \mathbb{R}^{m \times r}$, is called a *regular static
state-disturbance feedback* if $\beta(x)$ is invertible for all $x$. Note that the
feedback system (4.2.1,6) has the form

$$
(4.2.7) \quad \begin{cases} \dot{x} = (f+g\alpha)(x) + (g\beta)(x)v + (p+g\gamma)(x)q =: \tilde{f}(x) + \tilde{g}(x)v + \tilde{p}(x)q \\ y = h(x) \end{cases}
$$

Note that (4.2.7) is still affine in both $u$ and $q$. A smooth regular static state-disturbance feedback (4.2.6) is called a "friend" of a distribution $\Delta$, denoted as $(\alpha,\beta,\gamma) \in \mathcal{F}(\Delta)$, if

$$(4.2.8) \quad [\tilde{f},\Delta] \subset \Delta, \quad [\tilde{g}_i,\Delta] \subset \Delta, \quad i = 1,\ldots,m, \quad \tilde{p}_j \in \Delta, \quad j = 1,\ldots,r$$

**Definition 4.2.8**(a) *Local Disturbance Decoupling Problem with disturbance measurements* (LDDPdm) Consider the smooth nonlinear system (4.2.1) and a point $x_0 \in \mathbb{R}^n$. Under what conditions can we find a smooth regular static state-disturbance feedback (4.2.6) defined locally around $x_0$ such that in the feedback system (4.2.1,6) the disturbances $q$ do not influence the outputs $y$?

**Definition 4.2.8**(b) *Local Disturbance Decoupling Problem with Stability with disturbance measurements* (LDDPSdm) Consider the smooth nonlinear system (4.2.1). Under what conditions can we find a smooth regular static state-disturbance feedback (4.2.6) defined locally around $x = 0$ with $\alpha(0) = 0$ such that in the feedback system (4.2.1,6) the disturbances $q$ do not influence the outputs $y$ and $x = 0$ is a locally exponentially stable equilibrium of the modified drift dynamics $\dot{x} = f(x)+g(x)\alpha(x)$?

The solution of the LDDPdm is straightforward.

**Lemma 4.2.9** (cf. Theorem 2.3.8) Consider the smooth nonlinear system (4.2.1). Suppose that the distributions $\Delta^*$, $G$ and $\Delta^* + G$ are nonsingular. Then the Local Disturbance Decoupling Problem with disturbance measurements is solvable if and only if

$$(4.2.9) \quad P \subset \Delta^* + G$$

The proof is omitted (see e.g. [MG]).

Note that it follows from (4.2.9) that there exists a feedback (4.2.6) such that (4.2.8) holds for $\Delta^*$. To solve the LDDPSdm we apply Algorithm 4.2.1 with $P$ replaced by $\tilde{P} = \mathrm{sp}\{\tilde{p}_1,\ldots,\tilde{p}_r\}$. Suppose that the algorithm converges to $\Delta_\rho$ and let $\Delta^{\tilde{P}}_{\rho\,\tilde{f},\tilde{g},\tilde{p}} = \mathrm{inv\ clos}(\Delta_\rho)$. If $\Delta^{\tilde{P}}_{\rho\,\tilde{f},\tilde{g},\tilde{p}}$ is constant dimensional on $\mathcal{O}$ it does not depend on the feedback (4.2.6) with $(\alpha,\beta,\gamma) \in \mathcal{F}(\Delta^*)$. This immediately implies the following result.

**Theorem 4.2.10** Consider the smooth system (4.2.1). Assume that (4.2.9) holds and that (B2)-(B5) hold with $P$ replaced by $\tilde{P}$. Then the Local Disturbance Decoupling Problem with Stability with disturbance measurements for (4.2.1) is solvable.

Note that if (4.2.9) and (B2) up to (B5) hold, and if $\Delta_s^*$ is well-defined for the system (4.2.1), then solvability of the LDDPSdm implies that $\tilde{P} \subset (\Delta^{\tilde{P}})_* \subset \Delta_s^*$, so $P \subset \Delta_s^* + G$.

In the next section we comment on relations between the two methods to solve the LDDPS and the assumptions under which the distributions $\Delta_s^*$ and $(\Delta^P)_*$ are defined.

### 4.3  Comparison of the two methods to solve the LDDPS

In this section we take a closer look at the assumptions under which the distributions $\Delta_s^*$ and $(\Delta^P)_*$ are defined and the conditions under which the LDDPS is solvable for the smooth nonlinear system

$$(4.3.1) \quad \begin{cases} \dot{x} = f(x) + g(x)u + p(x)q, & f(0) = 0, \ x \in \mathbb{R}^n, \ u \in \mathbb{R}^m, \ q \in \mathbb{R}^r \\ y = h(x), & h(0) = 0, \ y \in \mathbb{R}^\ell \end{cases}$$

For convenience, we recall these assumptions here. For the solution of the LDDPS in terms of $\Delta_s^*$ we have (see Sections 3.2 and 3.3):

(A1)  $x = 0$ is a regular point of the decoupling matrix;

(A2)  $\Delta^{\tilde{f},\tilde{g}}$ has constant dimension on a neighborhood $\mathcal{O}$ of $x = 0$ and the integral manifold of $\Delta^{\tilde{f},\tilde{g}}$ through $x = 0$ is contained in the stable invariant manifold $S_0$;

(A3)  The system (4.3.1) is strongly accessible on $\mathcal{O}$;

(A4)  $\Pi^*$ has constant dimension and dim $G = m$ on $\mathcal{O}$;

(A5)  The linearization of the dynamics (4.3.1) restricted to the leaf $L_0$ of $\Pi^*$ through $x = 0$ is stabilizable;

(A6)   $\Delta^*$ and $\Delta^* + G$ have constant dimension on $\mathcal{O}$;

(A7)   The linearization of the system dynamics (4.3.1) modulo $\Delta_s^*$ is stabilizable.

For the solution of the LDDPS using $(\Delta^P)_*$ these assumptions are:

(B1)   $P \subset \Delta^*$ on $\mathcal{O}$;

(B2)   dim $G = m$ and $\Delta^*$ is constant dimensional on $\mathcal{O}$;

(B3)   $\Delta_{f,g}^P \sim$ has constant dimension on $\mathcal{O}$;

(B4)   The dynamics (4.3.1) restricted to the leaf of $(\Delta^P)_*$ through $x = 0$ can be stabilized exponentially;

(B5)   The linearization of the system dynamics (4.3.1) modulo $(\Delta^P)_*$ is stabilizable.

In Chapters 3 and 4 two essentially different methods are treated to find a solution for the LDDPS. In Chapter 3 we searched for the largest stabilizability distribution in the kernel of the output mapping, while in Section 4.2 we looked for the smallest controlled invariant distribution in ker $dh$ containing the disturbance vector fields and $\Pi^*$. While $\Delta_s^*$ is independent of the disturbance vector fields, $(\Delta^P)_*$ heavily depends on $P$. Therefore, the first method is more fundamental in contrast to the second method which is more "problem oriented". In order that $\Delta_s^*$ and $(\Delta^P)_*$ can be defined uniquely it is necessary that $\Pi^*$ is contained in these distributions. However, the role of the similar conditions (A5) and (B4) is totally different in both methods. While (A5) is essential in order that $\Pi^*$ is contained in $\Delta_s^*$, stabilizability of the restriction of the system dynamics to the leaf of $(\Delta^P)_*$ through the equilibrium ((B4)) is needed only for the solvability of the LDDPS, but not for the existence of $(\Delta^P)_*$.
A special role in Section 3.2 is played by assumption (A3). This condition is not included for the definition of $\Delta_s^*$, but to make sure that prolongation of the stable invariant manifold $S_0$ would lead to a foliation on $\mathbb{R}^n$. It follows immediately from the construction of this foliation that (A3) may be weakened to dim $(T_x S_0)$ + dim $C(x) = n$ for all $x$ in $\mathcal{O}$ (or to a similar condition for another stable invariant manifold in case prolongation of $S_0$ does not yield $\Delta_s^*$).

Assumption (A1) is contained in the list, because in Chapter 3 strongly input-output decouplable systems are considered first. It is easy to see that $A(x)$ can be written as

$$(4.3.2) \quad A(x) = \begin{bmatrix} dL_f^{r_1-1}h_1 \\ \vdots \\ dL_f^{r_\ell-1}h_\ell \end{bmatrix} (g_1, \ldots, g_m)$$

Hence, for square systems, (A1) implies (B2).

Apart from (A1) and (A3) the requirements in Chapters 3 and 4 are very much the same. Essentially these assumptions can be divided in two parts, viz. stabilizability and regularity conditions. Let us consider the stabilizability requirements first.

Recall from Remark 1.19 that a necessary condition for solvability of the LDDPS for (4.3.1) is that

(C1) The linearization of the system (4.3.1) around $x = 0$ is stabilizable.

Condition (C1) immediately implies (A7) and (B5) as follows from the following lemmas.

**Lemma 4.3.1** ([Tr]) Consider the stabilizable system $x = Ax + Bu$, $x \in \mathbb{R}^n$, $u \in \mathbb{R}^m$. Let $V$ denote a subspace in $\mathbb{R}^n$. Then the following two conditions are equivalent:

(i) $V$ is a stabilizability subspace.

(ii) There exists a feedback $u = Fx$ with $(A+BF)V \subset V$ such that $\sigma(A+BF) \subset \mathbb{C}^-$.

**Lemma 4.3.2** Consider the smooth nonlinear system

$$(4.3.3) \quad \dot{x} = f(x) + g(x)u, \qquad x \in \mathbb{R}^n, \ u \in \mathbb{R}^m, \ f(0) = 0$$

Assume that the linearization of this system is stabilizable and that $\Delta$ is a stabilizability distribution. Then $V \simeq \Delta(0)$ is a stabilizability subspace and the linearization of the dynamics (4.3.3) modulo $\Delta$ is stabilizable.

**Proof** Without loss of generality, we may assume that $\Delta = \mathrm{sp}\{\frac{\partial}{\partial x_1}\}$. Choose a smooth regular static state feedback

$$(4.3.4) \quad u = \alpha(x) + \beta(x)v, \qquad \alpha(0) = 0$$

with $(\alpha, \beta) \in \mathscr{F}(\Delta)$ such that the restriction of the modified drift dynamics $\dot{x} = (f+g\alpha)(x)$ to the leaf $S_0$ of $\Delta$ through $x = 0$ is exponentially stable. With $\tilde{f} := f + g\alpha$ and $\tilde{g}_i := (g\beta)_i$, $i = 1,\ldots,m$ the system (4.3.3,4) has the form

$$(4.3.5) \quad \begin{cases} \dot{x}_1 = \tilde{f}^1(x_1, x_2) + \tilde{g}^1(x_1, x_2)u \\ \dot{x}_2 = \tilde{f}^2(x_2) \quad\quad + \tilde{g}^2(x_2)u \end{cases}$$

The linearization of (4.3.5) around $x = 0$ is given by

$$(4.3.6) \quad \begin{cases} \dot{\tilde{x}}_1 = \tilde{A}_{11}\tilde{x}_1 + \tilde{A}_{12}\tilde{x}_2 + \tilde{B}_1\tilde{u} \\ \dot{\tilde{x}}_2 = \quad\quad\quad \tilde{A}_{22}\tilde{x}_2 + \tilde{B}_2\tilde{u} \end{cases}$$

with

$$(4.3.7) \quad \tilde{A} = \begin{bmatrix} \tilde{A}_{11} & \tilde{A}_{12} \\ 0 & \tilde{A}_{22} \end{bmatrix} = \frac{\partial\tilde{f}}{\partial x}(0), \quad \tilde{B} = \begin{bmatrix} \tilde{B}_1 \\ \tilde{B}_2 \end{bmatrix} = \tilde{g}(0)$$

Now $S_0 = \{x \in \mathbb{R}^n |\ x_2 = 0\}$, so the restriction of the modified drift dynamics to $S_0$ is equal to $\dot{\tilde{x}}_1 = \tilde{f}_1(\tilde{x}_1, 0)$. Hence, by assumption, $\sigma(\tilde{A}_{11}) \subset \mathbb{C}^-$. Since $\mathscr{V} = \mathrm{sp}\{e_1,\ldots,e_k\}$ with $k = \dim(x_1)$ and $\tilde{A} = \frac{\partial\tilde{f}}{\partial x}(0) = \frac{\partial f}{\partial x}(0)+g(0)\frac{\partial\alpha}{\partial x}(0)$, it follows from (4.3.6) that $\mathscr{V}$ is a stabilizability subspace (cf. Definition 1.6). Since $(\tilde{A}, \tilde{B})$ is stabilizable, it follows from Lemma 4.3.1 now that the pair $(\tilde{A}_{22}, \tilde{B}_2)$ is stabilizable, so the linearization of the dynamics (4.3.3) modulo $\Delta$ can be stabilized. $\quad\square$

The following well-known result is used in the next chapter. (Note that it is proved implicitly in the proof of Lemma 4.3.2.)

**Corollary 4.3.3** Consider the system (4.3.1). If $\Delta$ is a locally controlled invariant distribution, then $\mathscr{V} \simeq \Delta(0)$ is a controlled invariant subspace for the linearization of (4.3.1) around the equilibrium $x = 0$. Moreover, if $\Delta$ is invariant under $f$, then $\mathscr{V}$ is invariant under $\frac{\partial f}{\partial x}(0)$.

**Remark 4.3.4** The implication of (A7) and (B5) by (C1) also follows from the fact that the eigenvalues associated with the uncontrollable modes of the linearization of the system correspond necessarily to the eigenvalues of the linear approximation of the zero dynamics (see [BI3]). Hence, in order that the LDDPS is solvable, the uncontrollable modes are necessarily contained in $\Delta_s^*$ or $(\Delta^P)_*$ and these modes correspond to exponentially stable eigenvalues. □

The assumption that the system (4.3.1) can be stabilized exponentially does not make conditions like (A5) and (B4) superfluous, because it is possible that making $\Delta^*$ (and $\Pi^*$) invariant implies that some of the transmission zeros lie in the right-half plane. Moreover, it is possible that in that case the linearization of the dynamics restricted to $L_0$, the leaf of $\Pi^*$ through $x = 0$, is not stabilizable. (Recall from Section 2.3 that controllability distributions are not necessarily stabilizable.) These remarks are illustrated by the following example.

**Example 4.3.5** Consider the system (4.3.1) with $n = 4$, $m = 2$, $\ell = 1$ and

$$(4.3.8) \quad f(x) = \begin{pmatrix} -x_2 \\ 0 \\ x_3 + x_4 \\ 2x_4 \end{pmatrix}, \quad g(x) = \begin{pmatrix} 1 & 1 \\ x_2 & 1 \\ 0 & 0 \\ 0 & 3 \end{pmatrix}, \quad h(x) = x_4, \quad p(x) \text{ arbirary}$$

Since $L_{g_2} h(x) = 3$ for all $x$, we have $\Delta^* = \mathrm{sp}\{\frac{\partial}{\partial x_1}, \frac{\partial}{\partial x_2}, \frac{\partial}{\partial x_3}\}$. Moreover, $\Delta^* \cap G = \mathrm{sp}\{g_1\}$ and $\Pi^* = \mathrm{sp}\{\frac{\partial}{\partial x_1}, \frac{\partial}{\partial x_2}\}$. Now consider the linearization of (4.3.1,8) around $x = 0$:

$$(4.3.9) \quad \begin{cases} \dot{x} = Ax + Bu + Eq \\ y = Cx \end{cases}$$

with

$$(4.3.10) \quad A = \frac{\partial f}{\partial x}(0) = \begin{pmatrix} 0 & -1 & 0 & 0 \\ 0 & 0 & 0 & 0 \\ 0 & 0 & 1 & 1 \\ 0 & 0 & 0 & 2 \end{pmatrix}, \quad B = g(0) = \begin{pmatrix} 1 & 1 \\ 0 & 1 \\ 0 & 0 \\ 0 & 3 \end{pmatrix}, \quad C = (0\ 0\ 0\ 1)$$

Obviously, the linear system (4.3.9,10) is stabilizable (even controllable). However, the linearization of the system (4.3.1,8) restricted to the leaf $L_0$ of $\Pi^*$ through $x = 0$

$$(4.3.11) \quad \dot{x}_1 = -x_2 + u_1, \quad \dot{x}_2 = 0$$

is not stabilizable. Note that the transmission zero is equal to one (see (4.3.10)). □

We now turn to the regularity requirements. Although it is quite natural to require that dim $G = m$ (i.e. that all inputs are independent), it is not necessary as follows from Example 4.3.6. The other examples show that in order that the LDDPS is solvable, neither $\Pi^*$ nor $\Delta^*$ has to be constant dimensional.

**Example 4.3.6** Consider the linear system (4.3.1) with $n = 4$, $m = \ell = r = 1$ and

$$(4.3.12) \quad f(x) = -x, \quad g(x) = 0, \quad h(x) = x_4, \quad p(x) = \begin{pmatrix} 1 & 0 & 0 & 0 \end{pmatrix}^T$$

In this case, the DDPS is already solved, $V_s^* = V^* = \text{sp}\{e_1, e_2, e_3\}$. Note that the subspace $\hat{V}_E = \text{sp}\{e_1\} \not= V_s^*$, and that $\dim G = 0 \not= m$. □

**Example 4.3.7** Consider again the system (4.3.1) with $n = 4$, $m = 2$, $\ell = 1$ and $r = 1$. Let

$$(4.3.13) \quad f(x) = -x, \quad g(x) = \begin{bmatrix} x_3^2 & 0 \\ 0 & 1 \\ x_2 & 0 \\ 0 & 0 \end{bmatrix}, \quad p(x) = \begin{bmatrix} 1 \\ 0 \\ 0 \\ 0 \end{bmatrix}, \quad h(x) = x_2$$

Obviously, $\Delta^* = \text{sp}\{\frac{\partial}{\partial x_1}, \frac{\partial}{\partial x_3}, \frac{\partial}{\partial x_4}\}$, and, since $\dot{x} = f(x)$ is already exponentially stable, $\Delta_s^* = \Delta^*$. Moreover, $\Pi^* = \text{sp}\{g_1, \frac{\partial}{\partial x_3}\} = \text{sp}\{x_3^2 \frac{\partial}{\partial x_1}, \frac{\partial}{\partial x_3}\}$. Note that $\Pi^*$ is a singular distribution on $\mathcal{O}$, since $\dim \Pi^*(0) = 1$. So assumption (A4) is not satisfied. Nonetheless, $\Delta_s^*$ can be defined uniquely. (Also $(\Delta^P)_*$ can be calculated easily, $(\Delta^P)_* = \text{sp}\{\frac{\partial}{\partial x_1}, \frac{\partial}{\partial x_3}\}$.) □

**Example 4.3.8** Consider again the system (4.3.1) with $n = 4$, $m = \ell = r = 1$ and

$$(4.3.14) \quad f(x) = \begin{bmatrix} -x_1 \\ -x_2 \\ x_3 \\ -x_4 \end{bmatrix}, \quad g(x) = \begin{bmatrix} 0 \\ 0 \\ 1 \\ 0 \end{bmatrix}, \quad p(x) = \begin{bmatrix} 0 \\ 0 \\ 0 \\ 1 \end{bmatrix}, \quad h(x) = x_1 x_2$$

Since $\ker dh = \text{sp}\{x_1 \frac{\partial}{\partial x_1} - x_2 \frac{\partial}{\partial x_2}, \frac{\partial}{\partial x_3}, \frac{\partial}{\partial x_4}\}$ is invariant under $f$ and $g$ it is clear that $\Delta^*$ equals $\ker dh$. Hence, $\Delta^*$ is not a regular distribution. Obviously, $\Pi^* = \text{sp}\{\frac{\partial}{\partial x_3}\}$. By choosing

(4.3.15)  $u = -2x_3 + v$

the system (4.3.1,14,15) is exponentially stable. Consequently, the LDDPS for the system (4.3.1,14) can be solved by applying the feedback (4.3.15). Observe that in this case $(\Delta^P)_* = \mathrm{sp}\{\frac{\partial}{\partial x_3}, \frac{\partial}{\partial x_4}\}$ and that the feedback (4.3.15) stabilizes the dynamics restricted to the leaf of $(\Delta^P)_*$ through $x = 0$.  □

**Remark 4.3.9** It follows from the foregoing examples that the regularity conditions in Chapters 3 and 4 are in general not needed in order that the LDDPS is solvable. These requirements are imposed for convenience (as in much of the literature on nonlinear control theory). For the same reason the distributions $\Delta_s^*$ and $(\Delta^P)_*$ are explicitly required to be constant dimensional. (Note that a natural candidate for $\Delta_s^*$ in Example 4.3.8 would be the singular distribution $\Delta^*$.) It is clear from these considerations that a full solution of the LDDPS can be found only if singular distributions are taken into account.  □

## 4.4 The Disturbance Decoupling Problem with Stability in the literature

In this section we compare the results on the solution of the LDDPS we derived in Chapters 3 and 4 to the solution of the Disturbance Decoupling Problem with Bounded Disturbance Bounded State Stability (BDBS-DDP) given by Byrnes and Isidori in [BI2].
Consider the square smooth system (4.3.1). In [BI2] the BDBS-DDP is defined as follows:

**Definition 4.4.1** (BDBS-DDP) Consider the square smooth system (4.3.1). Let $\Omega$ be a bounded subset of $\mathbb{R}^n$ containing $x = 0$. Find - if possible - a globally defined feedback $u = \alpha(x)$ such that this feedback solves the DDP and

$$(4.4.1) \quad \forall \epsilon > 0 \quad \exists \delta > 0 \quad \forall x_0 \in \Omega \quad [(\,|q_j(t)| \le \delta, \ j = 1,\ldots,r)$$
$$\Rightarrow (\,|x(t)| < \epsilon, \ t \gg 0)]$$

**Remark 4.4.2** The problem defined here is a *global* problem contrary to the LDDPS we have considered. However, if we concentrate on the local aspects of the solution given in [BI2], then it appears that our solution is more general.  □

Byrnes and Isidori assume that (A1) and (B1) hold and that all vector fields are complete and that the disturbance vector fields are bounded. However, the two most important assumptions made in [BI2] are:

(H1)   The distribution $G$ is involutive;

(H2)   The zero dynamics of the system (4.3.1) are exponentially minimum phase.

In [BI2] it is suggested to achieve (H1) by adding integrators to (4.3.1) if (H1) does not hold. However, this yields a solution to the Disturbance Decoupling Problem by dynamic feedback, rather than by static feedback, see also Chapter 6.
It follows from the definition of exponentially minimum phase system (see Section 2.5) that assumption (H2) is very strong. Indeed, (H2) implies that $\Delta^*$ and $\Delta_s^*$ coincide.

Byrnes and Isidori show that, under the given conditions, the following holds:
For any bounded set $\Omega_r$ of initial data there exists a feedback law $u_r = \alpha_r(x)$ that solves the BDBS-DDP on $\Omega_r$.

Our assumptions are less restrictive, because we also allow for nonsquare systems and for systems that are not strongly input-output decouplable and/or exponentially minimum phase. Also, under the conditions in Chapter 3 we are able to solve the LDDPS which by Lemma 1.16 implies that the DDP with local BDBS-stability is solvable. However, as stressed before, the results in [BI2] are global, whereas our results are typically local.

## 5. Connections between the Solution of the LDDPS for a Nonlinear System and the DDPS for its Linearization

### 5.1 Introduction

In practical situations engineers often have to face difficult control problems for (highly) nonlinear systems. Instead of implementing advanced nonlinear control strategies (if these are available anyway) it is common practice to linearize the system around an equilibrium and to solve the control problem for this linearized system. The main idea behind this strategy is the following: "The linear feedback that solves the control problem for the linearized system at least approximately meets the control objectives if it is applied to the nonlinear system". Of course, this statement is not true in general. For some special design problems conditions have been found under which the statement holds true, see e.g. [HN] (where the design objective is model matching) and [GN] (strong input–output decoupling). In the next section we consider the relationship between solvability of the LDDPS for a nonlinear system and the DDPS for its linearization (see also [vdW2]). Moreover, we present a theorem on solvability of the LDDPS by applying a *linear* feedback.

### 5.2 Disturbance Decoupling with Stability for a nonlinear system and its linearization

In the foregoing chapters we gave two methods to solve the LDDPS for a smooth nonlinear system of the form

$$(5.2.1) \quad \begin{cases} \dot{x} = f(x) + g(x)u + p(x)q, \quad f(0) = 0, \ x \in \mathbb{R}^n, \ u \in \mathbb{R}^m, \ q \in \mathbb{R}^r \\ y = h(x), \qquad\qquad\qquad\qquad h(0) = 0, \ y \in \mathbb{R}^\ell \end{cases}$$

Also, the DDPS for the linearization of (5.2.1) around the equilibrium $x = 0$ given by

$$(5.2.2) \quad \begin{cases} \dot{\tilde{x}} = A\tilde{x} + B\tilde{u} + E\tilde{q} \\ \tilde{y} = C\tilde{x} \end{cases}$$

with

(5.2.3)    $A = \frac{\partial f}{\partial x}(0)$,    $B = g(0)$,    $E = p(0)$,    $C = \frac{\partial h}{\partial x}(0)$

was treated (in Chapter 1).

Now, some obvious questions arise. Is the LDDPS for (5.2.1) solvable if and only if the DDPS for (5.2.2) is? Is a feedback that solves the DDPS for (5.2.2) the linearization of a feedback that solves the LDDPS for (5.2.1)? These and related questions are addressed in this section. Note that solvability of the DDPS for (5.2.2) means that this problem can be solved by applying a *linear* regular static state feedback.

Recall from Remark 2.2.6 that if $\Delta$ is a distribution on $\mathbb{R}^n$, then $\Delta(0) \subset T_0\mathbb{R}^n$ can be identified with a subspace $V$ in $\mathbb{R}^n$ (identifying the tangent space $T_0\mathbb{R}^n$ with $\mathbb{R}^n$) and that the symbol $\Delta(0)$ is also used to denote this subspace $V$ in $\mathbb{R}^n$.


A partial answer to the questions raised above can be proved easily using Corollary 4.3.3. Solvability of the LDDPS for (5.2.1) implies that the DDPS for (5.2.2) can be solved. To be more specific, suppose that the smooth regular static state feedback

(5.2.4)    $u = \alpha(x) + \beta(x)v$,    $\alpha(0) = 0$

defined on a neighborhood $O$ of $x = 0$ solves the LDDPS for (5.2.1). Then the linearization

(5.2.5)    $\tilde{u} = F\tilde{x} + G\tilde{v}$,    $F = \frac{\partial \alpha}{\partial x}(0)$,    $G = \beta(0)$

solves the DDPS for (5.2.2).

The converse is in general not true as can be seen from the following simple example.


**Example 5.2.1**  Consider the system (5.2.1) with $n = 3$, $m = \ell = r = 1$ and

(5.2.6)    $f(x) = \begin{bmatrix} -x_1 \\ -x_2 \\ -2x_3 \end{bmatrix}$,    $g(x) = \begin{bmatrix} 1 \\ 0 \\ 1 \end{bmatrix}$,    $p(x) = \begin{bmatrix} 0 \\ 1 \\ x_2 \end{bmatrix}$,    $h(x) = x_3$

This implies that

(5.2.7)    $A = \begin{bmatrix} -1 & 0 & 0 \\ 0 & -1 & 0 \\ 0 & 0 & -2 \end{bmatrix}$,    $B = \begin{bmatrix} 1 \\ 0 \\ 1 \end{bmatrix}$,    $E = \begin{bmatrix} 0 \\ 1 \\ 0 \end{bmatrix}$,    $C = \begin{bmatrix} 0 & 0 & 1 \end{bmatrix}$

Now the DDPS for (5.2.2,7) is solvable, because im $E \subset V_s^* = V^* = \mathrm{sp}\{e_1, e_2\}$. On the other hand

$$(5.2.8) \qquad p(x) \notin \Delta^*(x) = \mathrm{sp}\{\frac{\partial}{\partial x_1}, \frac{\partial}{\partial x_2}\}, \qquad x \neq 0$$

so the LDDPS for (5.2.1,6) is not solvable around $x = 0$. $\qquad\qquad\qquad\square$

Note that things go wrong in the example, because the condition $p(0) \in \Delta^*(0)$ does not imply that $p \in \Delta^*$ for all $x$ in a neighborhood of $x = 0$. Thus in order that the LDDPS for the nonlinear system (5.2.1) is solvable it is necessary that, apart from solvability of the DDPS for (5.2.2), it is required explicitly that the LDDP for (5.2.1) is solvable. We assume in the sequel that

(D1) $P \subset \Delta^*$ on a neighborhood $\mathcal{O}$ of $x = 0$;

(D2) The equilibrium $x = 0$ is a regular point of the decoupling matrix.

Note that the conditions (D1) and (D2) are the same as (B1) and (A1) in the previous chapters. Recall that this implies that on $\mathcal{O}$ the relative degrees are finite and constant, say equal to $r_1, \ldots, r_\ell$, and that the decoupling matrix $A(x)$ defined by

$$(5.2.9) \qquad \left(A(x)\right)_{ij} = L_{g_j} L_f^{r_i - 1} h_i(x)$$

has full row rank.
It is proved in [GN] that in case (D2) holds, then the relative degrees of the linearization (5.2.2) are equal to $r_1, \ldots, r_\ell$ and $\Delta^*(0)$ can be identified with $V^*$. The main result in [GN] states that in case condition (D2) holds for the square system (5.2.1), then the Strong Local Input-Output Decoupling Problem for (5.2.1) is solvable if and only if it is solvable for (5.2.2). Note that (D2) holds for the system considered in Example 5.2.1. So, it is clear that (D2) is not sufficient in order that the LDDPS can be solved for (5.2.1) if and only if the DDPS can be solved for (5.2.2). Our main result (Theorem 5.2.2) gives conditions for the solvability of the LDDPS for (5.2.1) in terms of the linearization (5.2.2). A key role in this theorem is played by $(\Delta^P)_*$, the smallest constant dimensional locally controlled invariant distribution in ker $dh$ that contains the disturbance vector fields $P$ as well as $\Pi^*$, the largest local controllability distribution in ker $dh$. Assume that (see Section 4.2)

(D3)  $(\Delta^P)_*$ is uniquely defined on $\mathcal{O}$.

**Theorem 5.2.2** Consider the smooth nonlinear system (5.2.1) and its linearization (5.2.2) around the equilibrium $x = 0$. Assume that (D1), (D2) and (D3) hold and that

(D4)  dim $G = m$ and $\Pi^*$ is constant dimensional on $\mathcal{O}$;

(D5)  The linearization of the system (5.2.1) is stabilizable;

(D6)  The linearization of the dynamics (5.2.1) restricted to the leaf $L_0$ of $\Pi^*$ through $x = 0$ is stabilizable;

(D7)  $(\Delta^P)_*(0) \subset \mathcal{V}_s^*$.

Then the LDDPS for (5.2.1) is solvable.

**Proof**  Choose a regular static state feedback

(5.2.10)  $u = \alpha(x) + \beta(x)v$,    $\alpha(0) = 0$,    $(\alpha, \beta) \in \mathcal{F}(\Delta^*)$

defined on $\mathcal{O}$. Then $\Delta^*$, $(\Delta^P)_*$ and $\Pi^*$ are invariant under $\tilde{f} := f + g\alpha$ and $\tilde{g}_i := (g\beta)_i$, $i = 1, \ldots, m$ (see Algorithms 2.3.20 and 4.2.1). Since $\Pi^*$, $(\Delta^P)_*$ and $\Delta^*$ are constant dimensional and involutive and $\Pi^* \subset (\Delta^P)_* \subset \Delta^*$, it follows from Theorem 2.2.5 that there exists a coordinate transformation $z = \varphi(x)$ defined on a neighborhood $\tilde{\mathcal{O}} \subset \mathcal{O}$ of $x = 0$ such that

(5.2.11)  $\Pi^* = \mathrm{sp}\{\frac{\partial}{\partial z_1}\}$,    $(\Delta^P)_* = \mathrm{sp}\{\frac{\partial}{\partial z_1}, \frac{\partial}{\partial z_2}\}$,    $\Delta^* = \mathrm{sp}\{\frac{\partial}{\partial z_1}, \frac{\partial}{\partial z_2}, \frac{\partial}{\partial z_3}\}$

In these coordinates system (5.2.1,10) takes the form

(5.2.12)
$$
\begin{cases}
\dot{z}_1 = \tilde{f}^1(z_1, z_2, z_3, z_4) + \tilde{g}^{11}(z_1, z_2, z_3, z_4)v^1 + \tilde{g}^{12}(z_1, z_2, z_3, z_4)v^2 \\
\qquad\qquad\qquad\qquad\qquad\qquad\qquad\qquad + p^1(z_1, z_2, z_3, z_4)q \\[2mm]
\dot{z}_2 = \tilde{f}^2(z_2, z_3, z_4) \qquad\qquad\qquad + \tilde{g}^{22}(z_2, z_3, z_4)v^2 \\
\qquad\qquad\qquad\qquad\qquad\qquad\qquad\qquad + p^2(z_1, z_2, z_3, z_4)q \\[2mm]
\dot{z}_3 = \tilde{f}^3(z_3, z_4) \qquad\qquad\qquad\qquad + \tilde{g}^{32}(z_3, z_4)v^2 \\[2mm]
\dot{z}_4 = \tilde{f}^4(z_4) \qquad\qquad\qquad\qquad\qquad + \tilde{g}^{42}(z_4)v^2 \\[2mm]
y = h(z_4)
\end{cases}
$$

with

(5.2.13) $\quad G \cap \Delta^* = \mathrm{sp}\{\tilde{g}_1, \ldots, \tilde{g}_{m-\ell}\}, \quad v^1 = (v_1, \ldots, v_{m-\ell}), \quad v^2 = (v_{m-\ell+1}, \ldots, v_m)$

By (D6) the pair $\left(\dfrac{\partial \tilde{f}^1}{\partial z_1}(0), \tilde{g}^{11}(0)\right)$ is stabilizable. So, by applying a linear feedback

(5.2.14) $\quad v^1 = F_1 z_1 + w^1, \quad v^2 = w^2$

such that

(5.2.15) $\quad \sigma\left(\dfrac{\partial(\tilde{f}^1 + \tilde{g}^{11} F_1 z_1)}{\partial z_1}(0)\right) \subset \mathbb{C}^-$

the dynamics of the system (5.2.12,14) restricted to the leaf $L_0$ of $\Pi^*$ is exponentially stable. Moreover, the feedback (5.2.10,14) still leaves $(\Delta^P)_*$ and $\Delta^*$ invariant. The system (5.2.12,14) already fulfills the disturbance decoupling requirement. By (D7), the dynamics of the system (5.2.12,14) restricted to the leaf $S_0$ of $(\Delta^P)_*$ through $x = 0$ is exponentially stable. Finally, by (D5) and Lemma 4.3.2 it is possible to find a linear feedback

(5.2.16) $\quad w^1 = \hat{u}^1, \quad w^2 = F_3 z_3 + F_4 z_4 + \hat{u}^2$

such that the feedback (5.2.10,14,16) solves the LDDPS for (5.2.1). $\qquad \square$

The following corollary immediately follows from the proof of Theorem 5.2.2.

**Corollary 5.2.3** A linear feedback (5.2.5) that solves the DDPS for (5.2.2) is the linearization of some feedback (5.2.4) with $(\alpha, \beta) \in \mathcal{F}((\Delta^P)_*)$ that solves the LDDPS for (5.2.1) if and only if $F \in \mathcal{F}((\Delta^P)_*(0))$.

**Remark 5.2.4**
(i)   As a matter of fact, the assertion in Corollary 5.2.3 also holds if $(\Delta^P)_*$ is replaced by an arbitrary constant dimensional locally controlled invariant distribution $\Delta$ with $\mathrm{im}\, E \subset \Delta(0) \subset V_s^*$.
(ii)  Note that the only nonlinear part of the feedback (5.2.10,14,16) that solves the LDDPS is determined by (5.2.10), a feedback that can be chosen in such a way that it solves the SLIODP (see Section 2.3).
(iii) By comparison of Algorithms 1.10 and 2.3.20 it follows that $\mathcal{R}^* \subset \Pi^*(0)$. Assume that condition (D4) holds. Then (D6) $\Leftrightarrow \Pi^*(0) \subset V_s^*$.

Moreover, (D6) and (D7) together are equivalent to the condition that the linearization of the dynamics (5.2.1) restricted to the leaf $S_0$ of $(\Delta^P)_*$ through $x = 0$ is stabilizable (i.e. assumption (B4) in Section 4.3). (Note that (D4) is not necessary in order that (B4) holds, cf. Example 4.3.7). □

We give two examples now. The first one is meant to give a straightforward illustration of the theory. The planar two-link robot manipulator in Example 5.2.6 is considered in more detail. It illustrates Theorem 5.2.7 on the solution of the LDDPS by applying *linear* feedback.

**Example 5.2.5** Consider the system (5.2.1) with $n = 5$, $m = 2$, $r = \ell = 1$ and

$$f(x) = \begin{bmatrix} 2x_1+x_2+x_1x_2+x_1x_3 \\ -x_2 \\ -x_3+x_4x_5^2 \\ 3x_4+3x_5 \\ 4x_5 \end{bmatrix}, \quad g(x) = \begin{bmatrix} 1 & 0 \\ 0 & 0 \\ 1 & 0 \\ 0 & 0 \\ 0 & 1 \end{bmatrix}$$

(5.2.17)

$$p(x) = \begin{bmatrix} 1+x_3^2 \\ 1 \\ 0 \\ 0 \\ 0 \end{bmatrix}, \quad h(x) = x_5$$

Since $<dh,g_2>(x) = 1$ for all $x$, it follows that the relative degree equals 1 and that $A(x)$ has full row rank. Now, $\Delta^* = sp\{\frac{\partial}{\partial x_1},\frac{\partial}{\partial x_2},\frac{\partial}{\partial x_3},\frac{\partial}{\partial x_4}\}$ is invariant under $f$ and $g$ and contains $p$. Furthermore, $\Pi^* = sp\{\frac{\partial}{\partial x_1},\frac{\partial}{\partial x_3}\}$ and $(\Delta^P)_* = sp\{\frac{\partial}{\partial x_1},\frac{\partial}{\partial x_2},\frac{\partial}{\partial x_3}\}$ are constant dimensional and invariant under $f$ and $g$. The linearization of the system (5.2.1,17) is given by (5.2.2) with

(5.2.18) $A = \begin{bmatrix} 2 & 1 & 0 & 0 & 0 \\ 0 & -1 & 0 & 0 & 0 \\ 0 & 0 & -1 & 0 & 0 \\ 0 & 0 & 0 & 3 & 3 \\ 0 & 0 & 0 & 0 & 4 \end{bmatrix}$, $B = \begin{bmatrix} 1 & 0 \\ 0 & 0 \\ 1 & 0 \\ 0 & 0 \\ 0 & 1 \end{bmatrix}$, $E = \begin{bmatrix} 1 \\ 1 \\ 0 \\ 0 \\ 0 \end{bmatrix}$, $C = \begin{bmatrix} 0 & 0 & 0 & 0 & 1 \end{bmatrix}$

Note that $V_s^* = sp\{e_1,e_2,e_3\}$, so (D7) holds. By applying the feedback

(5.2.19) $u_1 = -10x_1+v_1$, $u_2 = v_2$

the drift dynamics of the system (5.2.1,17,19) restricted to the leaf $L_0$ of $\Pi^*$ through $x = 0$ have a locally exponentially stable equilibrium at $x = 0$. It can easily be seen that (D5) holds and that the feedback

(5.2.20)  $v_1 = w_1$,   $v_2 = -14x_4 - 14x_5 + w_2$

solves the LDDPS now.                                                            □

**Example 5.2.6**  In this example we consider a fully controllable planar two-link robot manipulator that is fixed on a rotating disc (see Figure 3).

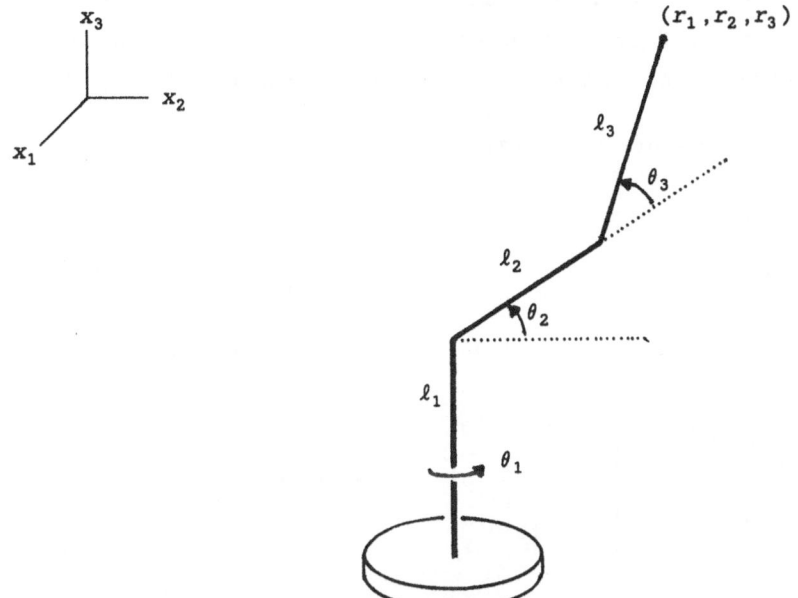

**Figure 3.** A two-link robot manipulator fixed on a rotating disc

This robot manipulator consists of three rigid links with length $\ell_1$, $\ell_2$ and $\ell_3$, respectively, with link 1 fixed on a rotating disc and links 2 and 3 moving in a plane. The configuration can be completely described by the following three angles:

$\theta_1$: the angle between the $x_1$-axis and the $(\ell_2, \ell_3)$-plane;

$\theta_2$: the angle between the $x_2$-axis and link 2;

$\theta_3$: the angle between links 2 and 3.

The coordinates $(r_1, r_2, r_3)$ of the endpoint are given by

$$(5.2.21) \quad \begin{cases} r_1 = [\ell_2\cos(\theta_2)+\ell_3\cos(\theta_2+\theta_3)]\cos(\theta_1) \\[4pt] r_2 = [\ell_2\cos(\theta_2)+\ell_3\cos(\theta_2+\theta_3)]\sin(\theta_1) \\[4pt] r_3 = \ell_1+\ell_2\sin(\theta_2)+\ell_3\sin(\theta_2+\theta_3) \end{cases}$$

We want to consider the LDDPS for this configuration in case the height is the output. To simplify the problem we first look at the robot arm equations without disturbances (cf. [Cr]). With $\theta = (\theta_1 \; \theta_2 \; \theta_3)^T$ the system equations are

$$(5.2.22) \quad M(\theta)\theta^{(2)} + B(\theta,\dot\theta) + k(\theta) = u$$

where $M(\theta)$ denotes the invertible mass matrix, $B(\theta,\dot\theta)$ the vector of Coriolis and centrifugal forces and $k(\theta)$ the vector containing the gravitation terms. Application of the state feedback (see [Cr])

$$(5.2.23) \quad u = M(\theta)v + B(\theta,\dot\theta) + k(\theta)$$

to (5.2.22) yields

$$(5.2.24) \quad \theta^{(2)} = v$$

Introducing $x = (\theta_1 \; \dot\theta_1 \; \theta_2 \; \dot\theta_2 \; \theta_3 \; \dot\theta_3)^T$ we can write (5.2.24) as

$$(5.2.25) \quad \dot{x} = \begin{bmatrix} 0 & 1 & 0 & 0 \\ 0 & 0 & & \\ & & 0 & 1 & 0 \\ 0 & & 0 & 0 & \\ & & & & 0 & 1 \\ 0 & & 0 & & 0 & 0 \end{bmatrix} x + \begin{bmatrix} 0 & 0 & 0 \\ 1 & & \\ & 0 & 0 & 0 \\ 0 & & 1 & \\ & & 0 & 0 & 0 \\ 0 & & 0 & & 1 \end{bmatrix} v =: Ax + Bv$$

The system we consider now is (5.2.25) where the state is influenced by some nonlinear disturbances together with the output $y = r_3-\ell_1$ with $r_3$ given by (5.2.21):

$$(5.2.26) \quad \begin{cases} \dot{x} = Ax + Bv + p(x)q =: f(x) + g(x)v + p(x)q \\[4pt] y = h(x) = \ell_2\sin(x_3)+\ell_3\sin(x_3+x_5) \end{cases}$$

where $A$ and $B$ are defined in (5.2.25) and $p(x)$ is arbitrary for the moment. Now $<dh, g_i>(x) = 0$ for $i = 1,2,3$, $<dL_f h, g_2>(x) = \ell_2\cos(x_3)+\ell_3\cos(x_3+x_5)$, so the relative degree equals 2 on a neighborhood $\mathcal{O}$ of $x = 0$. It follows that $\Delta^* = \ker dh \cap \ker dL_f h$, so

(5.2.27) $\Delta^* = \text{sp}\{ \frac{\partial}{\partial x_1}, \frac{\partial}{\partial x_2}, -\ell_3 \cos(x_3+x_5)\frac{\partial}{\partial x_4} +$

$\qquad + [\ell_2 \cos(x_3)+\ell_3 \cos(x_3+x_5)]\frac{\partial}{\partial x_6},$

$\qquad - \ell_3 \cos(x_3+x_5)\frac{\partial}{\partial x_3} + \ell_3(x_4+x_6)\sin(x_3+x_5)\frac{\partial}{\partial x_4} +$

$\qquad + [\ell_2 \cos(x_3)+\ell_3 \cos(x_3+x_5)]\frac{\partial}{\partial x_5} +$

$\qquad - [\ell_2 x_4 \sin(x_3)+\ell_3(x_4+x_6)\sin(x_3+x_5)]\frac{\partial}{\partial x_6} \}$

Furthermore, a feedback $v = \alpha(x)+\beta(x)\bar{v}$ that makes $\Delta^*$ invariant follows from the equations (cf. (2.3.11))

(5.2.28) $A(x)\alpha(x) + b(x) = 0, \qquad A(x)\beta(x) = \begin{pmatrix} 0 & 0 & 1 \end{pmatrix}$

where the decoupling matrix $A(x) = \begin{pmatrix} A_1(x) & A_2(x) & A_3(x) \end{pmatrix}$ is given by

(5.2.29) $A(x) = \begin{pmatrix} 0 & \ell_2 \cos(x_3)+\ell_3 \cos(x_3+x_5) & \ell_3 \cos(x_3+x_5) \end{pmatrix}$

and

(5.2.30) $b(x) = L_f^2 h(x) = -\ell_2 x_4^2 \sin(x_3)-\ell_3(x_4+x_6)^2 \sin(x_3+x_5)$

This yields e.g.

(5.2.31) $\alpha(x) = \begin{pmatrix} 0 & 0 & -b(x)/A_3(x) \end{pmatrix}, \qquad \beta(x) = \begin{bmatrix} 0 & 1 & 0 \\ -A_3(x) & 0 & 0 \\ A_2(x) & 0 & 1/A_3(x) \end{bmatrix}$

Obviously, the feedback is well-defined on a neighborhood $\mathcal{O}$ of $x = 0$. Suppose that $P \subset \Delta^*$, then conditions (B1) and (B2) in Section 4.2 are fulfilled. However, it is hard to find a coordinate transformation $z = \varphi(x)$ on $\mathcal{O}$ such that $\Delta^* = \text{sp}\{\frac{\partial}{\partial z_i}, i = 1,\ldots,4\}$. Consequently, it is very difficult to calculate $(\Delta^P)_*$ and the linearization of (5.2.26) restricted to the leaf $S_0$ of $(\Delta^P)_*$ through $x = 0$. Therefore, we consider the linearization of (5.2.26) around the equilibrium $x = 0$

(5.2.32) $\begin{cases} \dot{\tilde{x}} = A\tilde{x} + B\tilde{u} + E\tilde{q} \\ \tilde{y} = C\tilde{x} \end{cases}$

with

(5.2.33) $E = p(0), \qquad C = \frac{\partial h}{\partial x}(0) = \begin{pmatrix} 0 & 0 & \ell_2+\ell_3 & 0 & \ell_3 & 0 \end{pmatrix}$

Calculating $V^*$ for (5.2.32) yields

(5.2.34) $V^* = sp\{e_1, e_2, -\ell_3 e_3 + (\ell_2 + \ell_3) e_5, -\ell_3 e_4 + (\ell_2 + \ell_3) e_6\}$

Furthermore,

(5.2.35) $\mathcal{R}^* = V_s^* = sp\{e_1, e_2\}$

Note that the system (5.2.32) is stabilizable. It follows from condition
(D7) that the LDDPS for (5.2.26) is solvable only if $(\Delta^P)_*(0) \subset sp\{e_1, e_2\}$.
Suppose that the disturbance vector fields are

(5.2.36) $p_1(x) = \dfrac{\partial}{\partial x_1}, \quad p_2(x) = \dfrac{\partial}{\partial x_2}$

then it follows easily that $(\Delta^P)_* = sp\{\dfrac{\partial}{\partial x_1}, \dfrac{\partial}{\partial x_2}\}$. A feedback that solves the
LDDPS for (5.2.26,36) is e.g.

(5.2.37) $v_1 = -x_1 - 2x_2 + w_1, \quad v_2 = -x_3 - 2x_4 + w_2, \quad v_3 = -x_5 - 2x_6 + w_3$ □

Note that the feedback (5.2.37) that solves the LDDPS for the system
(5.2.26,36) is *linear*. This is no coincidence, but follows from the results
given below (see also [vdWN1]).
Consider the smooth system

(5.2.38) $\begin{cases} \dot{x} = Ax + Bu + Eq, \quad x \in \mathbb{R}^n, \ u \in \mathbb{R}^m, \ q \in \mathbb{R}^r \\ y = h(x), \qquad\qquad\quad y \in \mathbb{R}^\ell \end{cases}$

Then the linear system

(5.2.39) $\begin{cases} \dot{x} = Ax + Bu + Eq, \quad x \in \mathbb{R}^n, \ u \in \mathbb{R}^m, \ q \in \mathbb{R}^r \\ \bar{y} = Dx, \qquad\qquad\quad \bar{y} \in \mathbb{R}^k \end{cases}$

is called an *associated* system for the system (5.2.38) if

(5.2.40) $\ker D \subset \ker dh(x)$ for all $x$ in a neighborhood of $x = 0$

The following theorem holds.

**Theorem 5.2.7** The Local Disturbance Decoupling Problem with Stability for
the smooth system (5.2.38) is solvable by applying a linear regular static
state feedback if and only if there exists a matrix $D$ that fulfills

(5.2.40) and the Disturbance Decoupling Problem with Stability for the associated system (5.2.39) with this particular choice of $D$ is solvable.

**Proof**

($\Rightarrow$) Suppose that the LDDPS for (5.2.38) is solvable by applying the linear regular static state feedback

(5.2.41)   $u = Fx + Gv$

So, the outputs are decoupled from the disturbances for all initial conditions $x_0$ in a neigborhood $O$ of $x = 0$ and all controlled inputs $v$ as long as the state trajectories stay in $O$. Especially, for $x_0 = 0$ and $v = 0$ for the feedback system (5.2.38,41) holds that

(5.2.42)   $\displaystyle\int_0^t e^{(A+BF)(t-s)} Eq(s)ds \subset \ker\ dh(x)$

for all $t$ such that the integral is contained in $O$ and all $x$ in a neighborhood of $x_0 = 0$.
So the flat distribution $\Delta$ associated with the subspace $< A + BF|\ \text{im}\ E >$ is contained in $\ker\ dh$. Let $D$ be any matrix such that

(5.2.43)   $\ker\ D = < A + BF|\ \text{im}\ E >$

Now consider the associated linear system (5.2.39) with $D$ chosen as in (5.2.43). (Note that (5.2.40) holds by definition of $D$). Since $< A + BF|\ \text{im}\ E >$ is controlled invariant and $A + BF$ is exponentially stable it follows that for the linear system (5.2.39,43) $V_s^* = \ker\ D$ and, consequently, that the DDPS for (5.2.39,43) is solvable.
($\Leftarrow$) Suppose that the regular static state feedback (5.2.41) solves the DDPS for (5.2.39) with some $D$ that fulfills (5.2.40). Then the modified drift dynamics $\dot{x} = (A+BF)x$ are asymptotically stable and for all $t$

(5.2.44)   $\displaystyle\int_0^t e^{(A+BF)(t-s)} Eq(s)ds \subset \ker\ D \subset \ker\ dh$

so the disturbances $q$ do not influence the outputs $y$.
This completes the proof.                                                    □

**Corollary 5.2.8** Consider the smooth system

$$(5.2.45) \quad \begin{cases} \dot{x} = Ax + Bu + p(x)q, \ x \in \mathbb{R}^n, \ u \in \mathbb{R}^m, \ q \in \mathbb{R}^r \\ y = h(x), \ h(0) = 0, \ y \in \mathbb{R}^\ell \end{cases}$$

Then the Local Disturbance Decoupling Problem with Stability for (5.2.45) is solvable by applying a linear regular static state feedback if there exists a matrix $D$ that fulfills (5.2.40) such that for all $x$ the disturbance vector fields $p_i(x)$, $i = 1,\ldots,r$ are contained in the maximal stabilizability subspace $V_s^*$ in ker $D$ of the system (5.2.39) with that particular choice of $D$.

The proof follows immediately from Theorems 5.2.7 and 3.3.1 (see also Remark 3.3.3) and is omitted. Corollary 5.2.8 is very well fit to apply to state-space linearizable systems such as the robot arm considered in the previous example.

**Example 5.2.9** (cf. Example 5.2.6) Consider again the system (5.2.26). Assume that $P \subset \text{sp}\{\frac{\partial}{\partial x_1}, \frac{\partial}{\partial x_2}\}$. Note that

$$(5.2.46) \quad dh(x) = \begin{pmatrix} 0 & 0 & \ell_2\cos(x_3) + \ell_3\cos(x_3 + x_5) & 0 & \ell_3\cos(x_3 + x_5) & 0 \end{pmatrix}$$

and that

$$(5.2.47) \quad D = \begin{bmatrix} 0 & 0 & 1 & 0 & 0 & 0 \\ 0 & 0 & 0 & 0 & 1 & 0 \end{bmatrix}$$

fulfills (5.2.40). It is easy to verify that for the system

$$(5.2.48) \quad \begin{cases} \dot{x} = Ax + Bv + p(x)q \\ \bar{y} = Dx \end{cases}$$

$\Delta^* = \text{sp}\{\frac{\partial}{\partial x_1}, \frac{\partial}{\partial x_2}\}$. So the LDDPS for (5.2.48) and thus for (5.2.26) is solvable by applying the linear feedback (5.2.37). $\qquad \square$

# 6. THE LOCAL DYNAMIC DISTURBANCE DECOUPLING PROBLEM WITH STABILITY FOR NONLINEAR SYSTEMS

## 6.1 Introduction and problem formulation

In Chapter 1 the Local Disturbance Decoupling Problem with Stability (LDDPS) is defined as the problem of finding – for a given nonlinear system and equilibrium $x_0 = 0$ – conditions under which there exists a locally defined regular *static* state feedback such that for the feedback system the following holds:

(i)    The disturbances do not influence the outputs;

(ii)   $x_0 = 0$ is an exponentially stable equilibrium of the modified drift dynamics.

In Chapters 3 and 4 such conditions have been given. An obvious question now is whether or not the set of nonlinear systems for which (i) and (ii) can be accomplished becomes larger if we allow for *dynamic* state feedback, i.e. the conditions (i) and (ii) should hold for the feedback system obtained by applying dynamic state feedback to the given system.

We first consider the simpler question if there exist systems for which disturbance decoupling can be achieved by applying dynamic state feedback, but not by applying static state feedback. For linear systems the answer is negative. It has been proved in [Bh1] that if disturbance decoupling can be achieved by dynamic state feedback it can also be obtained by applying static state feedback. Next, consider the analytic nonlinear system

$$(6.1.1) \quad \begin{cases} \dot{x} = f(x) + g(x)u + p(x)q, & x \in \mathbb{R}^n, \ u \in \mathbb{R}^m, \ q \in \mathbb{R}^r \\ y = h(x), & y \in \mathbb{R}^\ell \end{cases}$$

Let $x_0$ be arbitrary. If (6.1.1) is a SISO system (i.e. $\ell = m = 1$), then the answer to the question raised above is also negative (at least if some additional assumptions hold). This can be seen in the following way. Let $r_u(x)$ denote the relative degree of the SISO system (6.1.1). Define analogously $r_q(x)$ as the smallest integer such that

$$(6.1.2) \quad \begin{cases} L_{p_j} L_f^k h(x) = 0 & j = 1, \ldots, r, \ k < r_q(x){-}1 \\ L_{p_j} L_f^{r_q(x)-1} h(x) \neq 0 & \text{for some } j \end{cases}$$

Assume that the integer $r_u(x)$ is finite and constant, say equal to $r_u$, and that $r_u < r_q(x)$ on a neighborhood $\mathscr{O}$ of $x_0$. Then

$$(6.1.3) \quad y^{(r_u)} = L_f^{(r_u)}h(x) + L_g L_f^{(r_u-1)}h(x)u + \sum_{i=1}^{r} L_{p_i} L_f^{(r_u-1)}h(x)q_i$$

$$= L_f^{(r_u)}h(x) + L_g L_f^{(r_u-1)}h(x)u$$

Since $L_g L_f^{(r_u-1)}h(x)$ is nonzero around $x_0$, it is clear from (6.1.3) that the feedback

$$(6.1.4) \quad u = \left(L_g L_f^{(r_u-1)}h(x)\right)^{-1}\left(-L_f^{(r_u)}h(x) + v\right)$$

implies $y^{(r_u)} = v$, so $y$ is decoupled from $q$ by applying this static state feedback. In the same way it can easily be seen that $r_u < r_q(x)$ is a necessary condition in order that there exists a regular static state feedback that accomplishes (i). Hence, for a SISO nonlinear system we have (under the given assumptions on $r_u(x)$) that the LDDP is locally solvable if and only if $r_q(x) > r_u$. It is obvious that by application of an analytic compensator

$$(6.1.5) \quad \begin{cases} \dot{z} = \alpha(x,z) + \beta(x,z)v, \ x \in \mathbb{R}^n, \ z \in \mathbb{R}^\mu \\ u = \gamma(x,z) + \delta(x,z)v, \ u \in \mathbb{R}, \ v \in \mathbb{R} \end{cases}$$

with input $v$, output $u$ and states $z$ to the SISO system (6.1.1) the relative degrees $r_v(x,z)$ and $r_q^e(x,z)$ do depend on $x$ and $z$. Suppose that $r_v(x,z)$ is finite and constant on a neighborhood of $(x_0,z_0)$, say equal to $r_v$, then the LDDP for (6.1.1,5) is solvable if (and only if) $r_v < r_q^e(x,z)$. It can easily be checked that it follows from the definition of the relative degrees $r_u(x)$, $r_q(x)$, $r_v(x,z)$ and $r_q^e(x,z)$ that the condition $r_v < r_q^e(x,z)$ implies that $r_u < r_q(x)$. Hence, if the output can be decoupled from the disturbances by applying dynamic state feedback (i.e. the LDDP is solvable for (6.1.1,5), this is possible by applying static state feedback too. By using similar arguments it follows that under regularity assumptions on the relative degrees analogous results hold for MISO systems and square MIMO systems. These regularity assumptions imply in particular that $x_0$ is a regular point of the decoupling matrix. If $x_0$ is *not* a regular point of the decoupling matrix, then the situation changes completely. Indeed, as can be seen from Example 6.1.1 something may be gained by applying dynamic state feedback.

**Example 6.1.1** Consider the system

$$(6.1.6) \quad \begin{cases} \dot{x}_1 = e^{x_2}u_1, \ \dot{x}_2 = x_5, \ \dot{x}_3 = x_2+x_4+e^{x_4}u_1, \ \dot{x}_4 = u_2, \ \dot{x}_5 = x_1u_1+q \\ y_1 = x_1, \qquad y_2 = x_3 \end{cases}$$

It follows from Algorithm 2.3.9 that $\Delta^* = 0$, so the LDDP is not solvable for any $x_0$. Note that $\dot{y}_1 = e^{x_2}u_1$, $\dot{y}_2 = x_2+x_4+e^{x_4}u_1$. Hence the decoupling matrix has rank 1 for all $x$. Now, apply an integrator to the first input, i.e.

$$(6.1.7) \quad u_1 = z, \quad \dot{z} = v_1, \quad u_2 = v_2$$

Then

$$(6.1.8) \quad \begin{cases} \dot{y}_1 = e^{x_2}z, & y_1^{(2)} = e^{x_2}x_5z+e^{x_2}v_1 \\ \dot{y}_2 = x_2+x_4+e^{x_4}z, & y_2^{(2)} = x_5+v_2+e^{x_4}(zv_2+v_1) \end{cases}$$

and the decoupling matrix $\begin{bmatrix} e^{x_2} & 0 \\ e^{x_4} & 1+e^{x_4}z \end{bmatrix}$ has full row rank if $1+e^{x_4}z \neq 0$. Application of the feedback

$$(6.1.9) \quad v_1 = -x_5z+e^{-x_2}\tilde{u}_1, \quad v_2 = (1+e^{x_4}z)^{-1}(-x_5-e^{x_4}v_1+\tilde{u}_2)$$

in a neighborhood of an initial point $(x_0,z_0)$ with $1+e^{x_4\,0}z_0 \neq 0$ yields for the system (6.1.6,7,9) $y_1^{(2)} = \tilde{u}_1$, $y_2^{(2)} = \tilde{u}_2$. So the compensator (6.1.7,9) with initial state $z_0 \neq -e^{x_4\,0}$ locally decouples the outputs from the disturbances around any point $x_0$. Since $z = u_1$, for $u_1$ the condition $u_1(0) \neq -e^{-x_4\,0}$ should hold. $\qquad\square$

The above remarks give rise to the definition of the Local Dynamic Disturbance Decoupling Problem (with Stability). First, we define what kind of compensators are allowed (see [dBGM]).

**Definition 6.1.2** An analytic compensator

$$(6.1.10) \quad \begin{cases} \dot{z} = \alpha(x,z) + \beta(x,z)v, \ x \in \mathbb{R}^n, \ z \in \mathbb{R}^\mu \\ u = \gamma(x,z) + \delta(x,z)v, \ u \in \mathbb{R}^m, \ v \in \mathbb{R}^m \end{cases}$$

is called a *regular dynamic state feedback* if the system

$$(6.1.11) \quad \begin{cases} \dot{x} = f(x) + g(x)u \\ \dot{z} = \alpha(x,z) + \beta(x,z)v, \ x \in \mathbb{R}^n, \ z \in \mathbb{R}^\mu \\ u = \gamma(x,z) + \delta(x,z)v, \ u \in \mathbb{R}^m, \ v \in \mathbb{R}^m \end{cases}$$

with inputs $v$ and outputs $u$ is invertible.

The definitions of left- and right-invertibility can be found e.g. in [Hil] and [Fl], [RN], respectively. For a square system (i.e. a system with the same number $m$ of inputs and outputs) like (6.1.11) the conditions for left- and right-invertibility are the same, viz. that the rank of the system (see Section 6.2) equals $m$. Therefore, in Definition 6.1.2 the term "invertible" is used. The invertibility assumption ensures that the number of effective control inputs before and after dynamic state feedback is the same (compare to the regularity of a static state feedback in Chapter 1).
In the rest of this chapter, we consider – contrary to the previous chapters – *analytic* systems. This means that also if it is not stated explicitly all distributions, feedbacks etc. are assumed to be analytic.

**Definition 6.1.3** *Local Dynamic Disturbance Decoupling Problem* (LDDDP) Consider the analytic system (6.1.1) and a point $x_0 \in \mathbb{R}^n$. Under what conditions can we find a point $z_0 \in \mathbb{R}^\mu$ and an analytic regular dynamic state feedback (6.1.10) locally defined around $(x_0, z_0)$ such that in the feedback system (6.1.1,10) the disturbances $q$ do not influence the outputs $y$?

**Definition 6.1.4** *Local Dynamic Disturbance Decoupling Problem with Stability* (LDDDPS) Consider the analytic system (6.1.1) with $f(0) = 0$. Under what conditions can we find an analytic regular dynamic state feedback (6.1.10) locally defined around $(x,z) = (0,0)$ such that in the feedback system (6.1.1,10) the disturbances $q$ do not influence the outputs $y$ and $(0,0)$ is a locally exponentially stable equilibrium of the modified drift dynamics $\dot{x} = f(x)+g(x)\gamma(x,z)$, $\dot{z} = \alpha(x,z)$?

In Sections 6.3 and 6.4 solutions for the LDDDP and the LDDDPS are given for square systems for which the Strong Local Dynamic Input-Output Decoupling Problem, abbreviated as SLDIODP, is solvable. The SLDIODP is defined as follows (see [DM], [NR], [XG]; cf. also Definition 2.3.15 for the SLIODP).

**Definition 6.1.5** *Strong Local Dynamic Input–Output Decoupling Problem* (SLDIODP) Consider the square analytic system (6.1.1) and a point $x_0 \in \mathbb{R}^n$. Under what conditions can we find a point $z_0 \in \mathbb{R}^\mu$ and an analytic regular dynamic state feedback (6.1.10) defined on a neighborhood of $(x_0, z_0)$ such that the feedback system (6.1.1,10) is strongly input–output decoupled around $(x_0, z_0)$?

The solvability of the SLDIODP can be checked by applying Singh's algorithm (or via the dynamic extension algorithm, see e.g. [NvdS4]). Moreover, if the SLDIODP is solvable, then a decoupling compensator follows immediately from the algorithm. These well–known topics (see e.g. [dBGM], [Mo], [Si2]) are treated in Section 6.2. The exposition in Sections 6.2 and 6.3 closely follows [HNvdW1] and [HNvdW3]. In [HNvdW2] an extension of the results in these sections to nonsquare and noninvertible systems can be found. The same topics are treated in a broader framework in the doctoral dissertation ([Hu]) of Huijberts that is published recently. Another (equivalent) solution for the LDDDP using a version of the dynamic extension algorithm can be found in [Res].

## 6.2  Singh's algorithm

In this section a modified version of Singh's algorithm is presented. Furthermore, it is shown that if the SLDIODP for a nonlinear system is solvable locally around a given point $x_0$, then a decoupling compensator can be found directly from Singh's algorithm. Finally, it is proved that if $x_0$ is a strongly regular point for Singh's algorithm (see Definition 6.2.2), then all such compensators are diffeomorphic (at least if the initial points $z_0$ are well chosen).

Singh's algorithm has been introduced in [Si1] for calculating the left–inverse of a nonlinear system, thereby generalizing the algorithm given in [Hi1]. The version given here is a reformulation of the one given in [dBGM] (see also [Hu], [HNvdW1]). In the sequel we restrict ourselves to square systems. Consider the square analytic system

$$(6.2.1) \quad \begin{cases} \dot{x} = f(x) + g(x)u, & x \in \mathbb{R}^n, \ u \in \mathbb{R}^m \\ y = h(x), & y \in \mathbb{R}^m \end{cases}$$

The rank of a matrix $M(x)$ over the field of meromorphic functions is denoted as rank $M(x)$, while $\text{rank}_{\mathbb{R}} M(x)$ denotes the rank of $M(x)$ over the real numbers. Recall that a meromorphic function is the quotient of two analytic functions.

**Algorithm 6.2.1** *Singh's algorithm*

Step 1

Calculate

$$(6.2.2) \quad \dot{y} = \frac{\partial h}{\partial x}\bigl(f(x) + g(x)u\bigr) =: a_1(x) + b_1(x)u$$

and define

$$(6.2.3) \quad \rho_1 := s_1 := \text{rank } b_1(x)$$

Permute, if necessary, the outputs in such a way that the first $\rho_1$ rows of $b_1(x)$ are linearly independent. Denote the first $\rho_1$ rows of $y$ by $\tilde{y}_1$, and the others by $\bar{y}_1$. Since the last rows of $b_1(x)$ are linearly dependent on the first $\rho_1$ rows, the equations (6.2.2) can be rewritten as

$$(6.2.4) \quad \begin{cases} \dot{\tilde{y}}_1 = \tilde{a}_1(x) + \tilde{b}_1(x)u \\ \dot{\bar{y}}_1 = \bar{y}_1(x, \dot{\tilde{y}}_1) \end{cases}$$

where the last equations are affine in $\dot{\tilde{y}}_1$. Finally, set $\tilde{B}_1(x) := \tilde{b}_1(x)$.

Step k+1

Suppose that in Steps 1 up to $k$ $\dot{\tilde{y}}_1, \ldots, \tilde{y}_k^{(k)}$ have been defined such that

$$(6.2.5) \quad \begin{cases} \dot{\tilde{y}}_1 = \tilde{a}_1(x) + \tilde{b}_1(x)u \\ \quad \vdots \\ \tilde{y}_k^{(k)} = \tilde{a}_k\bigl(x, \{\tilde{y}_i^{(j)} \mid 1 \le i \le k-1, \ i \le j \le k\}\bigr) \\ \qquad + \tilde{b}_k\bigl(x, \{\tilde{y}_i^{(j)} \mid 1 \le i \le k-1, \ i \le j \le k-1\}\bigr)u \\ \bar{y}_k^{(k)} = \bar{y}_k^{(k)}\bigl(x, \{\tilde{y}_i^{(j)} \mid 1 \le i \le k, \ i \le j \le k\}\bigr) \end{cases}$$

and that the matrix $\tilde{B}_k = \left(\tilde{b}_1^T, \ldots, \tilde{b}_k^T\right)^T$ has full rank equal to $\rho_k$. Calculate

$$(6.2.6) \quad \bar{y}_k^{(k+1)} = \frac{\partial}{\partial x} \bar{y}_k^{(k)}(f(x) + g(x)u) + \sum_{i=1}^{k} \sum_{j=i}^{k} \frac{\partial \bar{y}_k^{(k)}}{\partial \tilde{y}_i^{(j)}} \tilde{y}_i^{(j+1)}$$

and write it as

$$(6.2.7) \quad \bar{y}_k^{(k+1)} = a_{k+1}\left(x, \{\tilde{y}_i^{(j)} \mid 1 \le i \le k, \ i \le j \le k+1\}\right) +$$

$$b_{k+1}\left(x, \{\tilde{y}_i^{(j)} \mid 1 \le i \le k, \ i \le j \le k\}\right)u$$

Define $B_{k+1} := \left(\tilde{B}_k^T, b_{k+1}^T\right)^T$ and $\rho_{k+1} := \text{rank } B_{k+1}$. Permute, if necessary, the components of $\bar{y}_k^{(k+1)}$ in such a way that the first $\rho_{k+1}$ rows of $B_{k+1}$ are linearly independent. Let $\tilde{y}_{k+1}^{(k+1)}$ denote the first $s_{k+1} := (\rho_{k+1} - \rho_k)$ rows of $\bar{y}_k^{(k+1)}$ and $\bar{y}_{k+1}^{(k+1)}$ the remaining rows. Since the last rows of $B_{k+1}$ are linearly dependent on the first $\rho_{k+1}$ rows, we can write the equations (6.2.5,7) as

$$(6.2.8) \quad \begin{cases} \dot{\tilde{y}}_1 = \tilde{a}_1(x) + \tilde{b}_1(x)u \\ \quad \vdots \\ \tilde{y}_{k+1}^{(k+1)} = \tilde{a}_{k+1}\left(x, \{\tilde{y}_i^{(j)} \mid 1 \le i \le k, \ i \le j \le k+1\}\right) \\ \qquad + \tilde{b}_{k+1}\left(x, \{\tilde{y}_i^{(j)} \mid 1 \le i \le k, \ i \le j \le k\}\right)u \\ \bar{y}_{k+1}^{(k+1)} = \bar{y}_{k+1}^{(k+1)}\left(x, \{\tilde{y}_i^{(j)} \mid 1 \le i \le k+1, \ i \le j \le k+1\}\right) \end{cases}$$

Finally, set $\tilde{B}_{k+1} := \left(\tilde{B}_k^T, \tilde{b}_{k+1}^T\right)^T$.  □

It is shown in [Mo] that the integers $\rho_1, \rho_2, \ldots$ do not depend on the particular permutation of the rows of $\bar{y}_k^{(k+1)}$. Thus, using Algorithm 6.2.1 we obtain a uniquely defined sequence of integers $0 \le \rho_1 \le \cdots \le \rho_k \le \cdots$ $\cdots \le m$. Let $\rho^* := \max\{\rho_k \mid k \ge 1\}$ and let $\alpha$ be the smallest integer such that $\rho_\alpha = \rho^*$. Then $\alpha \le n$. The number $\rho^*$ is called the *rank* of the system (6.2.1).

In the sequel we use the following notion of regularity associated with Singh's algorithm.

**Definition 6.2.2** Consider the square analytic system (6.2.1) and a point $x_0 \in \mathbb{R}^n$. We call $x_0$ a *regular point for Singh's algorithm* if for an appropriate application of the algorithm there exist $\tilde{y}_{i0}^{(j)}$ ($1 \leq i \leq n-1$, $i \leq j \leq n-1$) such that for each $1 \leq k \leq n$

$$(6.2.9) \quad \text{rank } B_k\left(x, \{\tilde{y}_i^{(j)} \mid 1 \leq i \leq k-1, \ i \leq j \leq k-1\}\right)$$

$$= \text{rank}_{\mathbb{R}} B_k\left(x_0, \{\tilde{y}_{i0}^{(j)} \mid 1 \leq i \leq k-1, \ i \leq j \leq k-1\}\right)$$

and

$$(6.2.10) \quad \text{rank} \begin{bmatrix} \dfrac{\partial h}{\partial x}(x) \\[2ex] \dfrac{\partial \bar{y}_1^{(1)}}{\partial x}(x, \tilde{y}_1^{(1)}) \\[1ex] \vdots \\[1ex] \dfrac{\partial \bar{y}_k^{(k)}}{\partial x}\left(x, \{\tilde{y}_i^{(j)} \mid 1 \leq i \leq k, \ i \leq j \leq k\}\right) \end{bmatrix}$$

$$= \text{rank}_{\mathbb{R}} \begin{bmatrix} \dfrac{\partial h}{\partial x}(x_0) \\[2ex] \dfrac{\partial \bar{y}_1^{(1)}}{\partial x}(x_0, \tilde{y}_{10}^{(1)}) \\[1ex] \vdots \\[1ex] \dfrac{\partial \bar{y}_k^{(k)}}{\partial x}\left(x_0, \{\tilde{y}_{i0}^{(j)} \mid 1 \leq i \leq k, \ i \leq j \leq k\}\right) \end{bmatrix}$$

We call $x_0$ a *strongly regular point for Singh's algorithm* if for each application of the algorithm there exist $\tilde{y}_{i0}^{(j)}$ ($1 \leq i \leq n-1$, $i \leq j \leq n-1$) such that, for each $1 \leq k \leq n$, the conditions (6.2.9) and (6.2.10) hold.

**Remark 6.2.3** Di Benedetto and Grizzle (see [dBG]) give another definition of regularity that takes explicitly into account that the $\tilde{y}_{i0}^{(j)}$ ($1 \leq i \leq n-1$, $i \leq j \leq n-1$) are realizable, i.e. that there exists a control $u$ such that

$$(6.2.11) \quad \tilde{y}_{i0}^{(j)} = \tilde{y}_i^{(j)}(u_0, \ldots, u_0^{(n-1)}), \quad 1 \leq i \leq n-1, \ i \leq j \leq n-1$$

It is shown by Huijberts (see [Hu]) that if $x_0$ is a regular point for Singh's algorithm, then there exists a control $u$ such that (6.2.11) holds. This implies that the approaches of regularity here (and in e.g. [Hu]) and in [dBG] in fact show two sides of the same problem, namely the output-side and the input-side, respectively.                                    □

It is well-known from the literature that if $x_0$ is a regular point for Singh's algorithm, then the SLDIODP for (6.2.1) is solvable if and only if $\rho^* = m$, which on its turn is equivalent to right-invertibility of the system (6.2.1) (see e.g. [dBGM], [DM], [RN], [Si2]). Assume that $x_0$ is a regular point for Singh's algorithm and that $\rho^* = m$. Then a regular dynamic state feedback that solves the SLDIODP can be found as follows. For an appropriate application of Singh's algorithm we have at the $\alpha$-th step:

(6.2.12)  $\dot{\tilde{Y}}_\alpha = \tilde{A}_\alpha\bigl(x, \{\tilde{y}_i^{(j)}| \ 1 \le i \le \alpha-1, \ i \le j \le \alpha\}\bigr)$

$\qquad\qquad + \tilde{B}_\alpha\bigl(x, \{\tilde{y}_i^{(j)}| \ 1 \le i \le \alpha-1, \ i \le j \le \alpha-1\}\bigr)u$

where $\tilde{Y}_\alpha = \bigl(\tilde{y}_1^T, \ldots, \tilde{y}_\alpha^{(\alpha-1)T}\bigr)^T$ and $\tilde{B}_\alpha\bigl(x, \{\tilde{y}_i^{(j)}| \ 1 \le i \le \alpha-1, \ i \le j \le \alpha-1\}\bigr)$ has full rank $m$. From (6.2.12) we obtain

(6.2.13)  $u = \tilde{B}_\alpha^{-1}\bigl(\dot{\tilde{Y}}_\alpha - \tilde{A}_\alpha\bigr) =: \varphi\bigl(x, \{\tilde{y}_i^{(j)}| \ 1 \le i \le \alpha, \ i \le j \le \alpha\}\bigr)$

Let $\gamma_i$ be the lowest time-derivative and $\delta_i$ the highest time-derivative of $\tilde{y}_i$ $(i = 1, \ldots, m)$ appearing in (6.2.13). Then it can be quite easily shown that $\varphi$ is of the form

(6.2.14)  $\varphi\bigl(x, \{\tilde{y}_i^{(j)}| \ 1 \le i \le m, \ \gamma_i \le j \le \delta_i\}\bigr)$

$\qquad = \varphi_1\bigl(x, \{\tilde{y}_i^{(j)}| \ 1 \le i \le m, \ \gamma_i \le j \le \delta_i-1\}\bigr)$

$\qquad\qquad + \sum_{i=1}^{m} \varphi_{2i}\bigl(x, \{\tilde{y}_i^{(j)}| \ 1 \le i \le m, \ \gamma_i \le j \le \delta_i-1\}\bigr)\tilde{y}_i^{(\delta_i)}$

Now let $z_i$ $(i = 1, \ldots, m)$ be a vector of dimension $\delta_i - \gamma_i$ and consider the system

(6.2.15)  $\begin{cases} \dot{z}_i = A_i z_i + B_i v_i, \qquad i = 1, \ldots, m \\[2mm] u = \varphi_1(x, z_1, \ldots, z_m) + \displaystyle\sum_{i=1}^{m} \varphi_{2i}(x, z_1, \ldots, z_m)v_i \end{cases}$

with inputs $v$, outputs $u$ and

$$(6.2.16) \quad A_i = \begin{bmatrix} 0 & 1 & & 0 \\ & \ddots & \ddots & \\ & & & 1 \\ 0 & & & 0 \end{bmatrix}_{(\delta_i - \gamma_i) \times (\delta_i - \gamma_i)} \qquad B_i = \begin{bmatrix} 0 \\ \vdots \\ 0 \\ 1 \end{bmatrix}_{\delta_i - \gamma_i} \quad , \quad i = 1, \ldots, m$$

Comparison of (6.2.13,14) and (6.2.15,16) immediately yields $\tilde{y}_i^{(\delta_i)} = v_i$, $i = 1, \ldots, m$, so the SLDIODP for (6.2.1) can be solved by applying a compensator of the form (6.2.15,16) (or (6.2.15) in short if no confusion is possible). It is proved in [Hu] that the compensator (6.2.15,16) is a regular dynamic state feedback.

Note that the point $z_0 \in \mathbb{R}^\mu$ can be taken as part of the $\tilde{y}_{i0}^{(j)}$ that fulfill (6.2.9) and (6.2.10). Since a compensator (6.2.15) is often used in the sequel, we refer to it as a Singh compensator.

**Definition 6.2.4** Consider the square analytic system (6.2.1). Any analytic regular dynamic state feedback (6.2.15,16) obtained by some application of Singh's algorithm is called a *Singh compensator*.

Obviously, the Singh compensator (6.2.15) depends on the permutation of the outputs at each step in the algorithm. However, the constants $\gamma_i$ and $\delta_i$, $i = 1, \ldots, m$, are *intrinsic*, i.e. $\{(\gamma_1, \delta_1), \ldots, (\gamma_m, \delta_m)\}$ is an invariant set for the system (6.2.1) (and thus independent of the permutation of the outputs at each step in an arbitrary application of Singh's algorithm). This implies that two arbitrary Singh compensators are strongly related.

**Proposition 6.2.5** Consider the square analytic system (6.2.1). Assume that $x_0$ is a strongly regular point for Singh's algorithm. Then the constants $\gamma_i$ and $\delta_i$ ($i = 1, \ldots, m$) are intrinsic. Suppose that (6.2.15,16) and

$$(6.2.17) \quad \begin{cases} \dot{z}_i = \bar{A}_i \bar{z}_i + \bar{B}_i v_i & , \quad i = 1, \ldots, m \\ u = \bar{\varphi}_1(x, \bar{z}_1, \ldots, \bar{z}_m) + \sum_{i=1}^{m} \bar{\varphi}_{2i}(x, \bar{z}_1, \ldots, \bar{z}_m) v_i \end{cases}$$

with

$$(6.2.18) \quad \bar{A}_i = \begin{bmatrix} 0 & 1 & & 0 \\ & \ddots & \ddots & \\ & & & 1 \\ 0 & & & 0 \end{bmatrix}_{(\bar{\delta}_i - \bar{\gamma}_i) \times (\bar{\delta}_i - \bar{\gamma}_i)} \qquad \bar{B}_i = \begin{bmatrix} 0 \\ \vdots \\ 0 \\ 1 \end{bmatrix}_{\bar{\delta}_i - \bar{\gamma}_i} \quad , \quad i = 1, \ldots, m$$

denote two arbitrary Singh compensators obtained for the system (6.2.1), then there exist initial points $z_0$ and $\bar{z}_0$ and an analytic coordinate transformation $(x,\bar{z}) = \Phi(x,z)$ defined locally around $(x_0,z_0)$ transforming the compensator (6.2.15,16) in (6.2.17,18).

The proof of this proposition can be found in Appendix B. The diffeomorphism $\Phi$ has the property $\frac{\partial\Phi}{\partial x}(x,z) = I_n$, so there also exists an $x$-dependent diffeomorphism $\bar{z} = \bar{\Phi}(x,z)$ transforming one compensator in the other.

**Remark 6.2.6** In [HNvdW1], [HNvdW2] and [HNvdW3] a compensator (6.1.10) is said to be a regular dynamic state feedback if the system (6.1.10) with inputs $v$ and outputs $u$ is invertible for all $z$ and all constant $x$. In case (6.1.10) is regular in this sense then the feedback system (6.2.1,1.10) has the same rank as the original system (6.2.1). If a compensator (6.1.10) is regular in the sense of Definition 6.1.2 then the precompensated system has this rank preservation property as well as an invertibility property (see [dBGM] and [Hu]). Singh compensators are regular in both senses, but it is not completely clear whether or not these definitions of regularity are equal for general compensators of the form (6.1.10). □

**Remark 6.2.7** In [HNvdW4] it is shown that the regular dynamic state feedback (6.2.15) is in fact a minimal order decoupling compensator for the system (6.2.1), i.e. the dimension $\mu = \sum_{i=1}^{m} (\delta_i - \gamma_i)$ of the state-space of the system (6.2.15) is minimal. This result extends earlier results on minimal order decoupling (see e.g. [GM],[XG]). □

Note that if $x_0$ is a strongly regular point for Singh's algorithm, then any application of the algorithm yields a compensator that solves the SLDIODP. It is shown in the next section that if the LDDDP is solvable for a square system for which the SLDIODP can be solved, then an arbitrary application of Singh's algorithm (i.e. no matter what permutation of the outputs is taken at each step) provides a compensator that solves the LDDDP.
This strong regularity assumption is imposed for convenience. In fact regularity is sufficient, because if the SLDIODP is solvable, then *one* particular application of Singh's algorithm is enough to find a decoupling compensator. Indeed, it follows from Proposition 6.2.5 that in a sense all Singh compensators are equivalent.

## 6.3 The Local Dynamic Disturbance Decoupling Problem

Having introduced Singh compensators in the previous section, a solution for the LDDDP for systems that are strongly dynamically input–output decouplable can easily be obtained now. Consider the square analytic system

$$(6.3.1) \quad \begin{cases} \dot{x} = f(x) + g(x)u + p(x)q, \ x \in \mathbb{R}^n, \ u \in \mathbb{R}^m, \ q \in \mathbb{R}^r \\ y = h(x), \qquad\qquad\qquad y \in \mathbb{R}^m \end{cases}$$

We say that the SLDIODP for (6.3.1) is solvable, if it can be solved for the system (6.3.1) with $q = 0$. Our main result is stated in the following theorem.

**Theorem 6.3.1** Consider the square analytic system (6.3.1) and a point $x_0 \in \mathbb{R}^n$. Assume that $x_0$ is a strongly regular point for Singh's algorithm and that the Strong Local Dynamic Input–Output Decoupling Problem for (6.3.1) is solvable. Then the Local Dynamic Disturbance Decoupling Problem is solvable if and only if it is solvable by applying a compensator (6.2.15) that is obtained by an arbitrary application of Singh's algorithm.

**Proof** The sufficiency part is trivial. Now assume that the LDDDP is solvable by the regular dynamic state feedback

$$(6.3.2) \quad \begin{cases} \dot{z} = \alpha(x,z) + \beta(x,z)v, \ x \in \mathbb{R}^n, \ z \in \mathbb{R}^\mu \\ u = \gamma(x,z) + \delta(x,z)v, \ u \in \mathbb{R}^m, \ v \in \mathbb{R}^m \end{cases}$$

defined locally around $(x_0, z_0)$ for some $z_0 \in \mathbb{R}^\mu$. If we consider the disturbances $q$ in (6.3.1) as parameters the first step of Singh's algorithm applied to (6.3.1) yields

$$(6.3.3) \quad \begin{cases} \dot{\tilde{y}}_1 = \tilde{a}_1(x,q) + \tilde{b}_1(x)u \\ \dot{\tilde{y}}_1 = \dot{\tilde{y}}_1(x,q,\dot{\tilde{y}}_1) \end{cases}$$

If we substitute the outputs $u$ of the compensator (6.3.2) in (6.3.3), the equations (6.3.3,2) have the form

$$(6.3.4) \quad \begin{cases} \dot{\tilde{y}}_1 = \tilde{a}_1(x,q) + \tilde{b}_1(x)\big(\gamma(x,z) + \delta(x,z)v\big) \\[2mm] \dot{\bar{y}}_1 = \bar{y}_1\big(x,q,\tilde{a}_1(x,q) + \tilde{b}_1(x)(\gamma(x,z) + \delta(x,z)v)\big) \end{cases}$$

Since (6.3.2) solves the LDDDP for (6.3.1) the $\dot{\bar{y}}_1$ and $\dot{\tilde{y}}_1$ in (6.3.4) must be independent of $q$. Because $\gamma(x,z) + \delta(x,z)v$ does not depend on $q$, it is obvious that $\dfrac{\partial \tilde{a}_1}{\partial q} = 0$ and so, $\dot{\tilde{y}}_1 = \bar{y}_1\big(x,q,\tilde{a}_1(x) + \tilde{b}_1(x)(\gamma(x,z) + \delta(x,z)v)\big)$. Furthermore, also $\dot{\tilde{y}}_1$ is independent of $q$, so $\dot{\bar{y}}_1 = \bar{y}_1(x,\dot{\tilde{y}}_1)$. Repeating this argument Singh's algorithm yields exactly the equations (6.2.12), so any Singh compensator (6.2.15) solves the LDDDP for (6.3.1). □

**Remark 6.3.2** Recall from Chapter 2 that if $x_0$ is a regular point of the decoupling matrix and the LDDP is solvable, then the LDDP can be solved by applying a regular static state feedback that solves the SLIODP. Theorem 6.3.1 generalizes this result to systems that are strongly dynamically input–output decouplable. □

Before giving geometric conditions for solvability of the LDDDP we consider Example 6.1.1 again.

**Example 6.3.3** (cf. Example 6.1.1) Consider the system

$$(6.3.5) \quad \begin{cases} \dot{x}_1 = e^{x_2}u_1, \ \dot{x}_2 = x_5, \ \dot{x}_3 = x_2 + x_4 + e^{x_4}u_1, \ \dot{x}_4 = u_2, \ \dot{x}_5 = x_1u_1 + q \\[2mm] y_1 = x_1, \qquad y_2 = x_3 \end{cases}$$

Apply Singh's algorithm to (6.3.5). This yields

$$(6.3.6) \quad \begin{cases} \dot{\tilde{y}}_1 = \dot{y}_1 = e^{x_2}u_1 \\[2mm] \dot{\tilde{y}}_1 = \dot{y}_2 = x_2 + x_4 + e^{x_4}u_1 = x_2 + x_4 + e^{x_4 - x_2}\dot{\tilde{y}}_1 \end{cases}$$

and

$$(6.3.7) \quad \bar{y}_1^{(2)} = x_5 + u_2 + e^{x_4 - x_2}(u_2\dot{\tilde{y}}_1 - x_5\dot{\tilde{y}}_1) + e^{x_4 - x_2}\tilde{y}_1^{(2)}$$

Hence

$$(6.3.8) \quad \begin{cases} \dot{y}_1 = e^{x_2}u_1 \\[2mm] y_2^{(2)} = x_5 + u_2 + e^{x_4 - x_2}\big((u_2 - x_5)\dot{y}_1 + y_1^{(2)}\big) \end{cases}$$

So $\gamma_1 = 1$, $\delta_1 = 2$ and $\gamma_2 = \delta_2 = 2$. Thus setting $\dot{y}_1 = z$, $y_1^{(2)} = v_1$ and $y_2^{(2)} = v_2$ we obtain the following compensator solving the LDDDP for (6.3.5) locally around any point $(x_0, z_0)$ with $1 + z_0 e^{x_{40} - x_{20}} \neq 0$:

$$(6.3.9) \quad \begin{cases} \dot{z} = v_1 \\ u_1 = e^{-x_2} z \\ u_2 = (1 + z e^{x_4 - x_2})^{-1}(-x_5 + x_5 z e^{x_4 - x_2} - e^{x_4 - x_2} v_1 + v_2) \end{cases}$$

Obviously, the largest locally controlled invariant distribution $\Delta_\bullet^*$ in ker $dh$ of the compensated system (6.3.5,9) depends on $x$ and $z$. $\Delta_\bullet^*$ is given by

$$(6.3.10) \quad \Delta_\bullet^* = \ker dx_1 \cap \ker dx_3 \cap \ker dz \cap \ker d(x_2 + x_4 + e^{x_4 - x_2} z)$$

$$= \mathrm{sp}\{\frac{\partial}{\partial x_5}, (1 + z e^{x_4 - x_2})\frac{\partial}{\partial x_2} - (1 - z e^{x_4 - x_2})\frac{\partial}{\partial x_4}\}$$

Note that dim $\Delta_\bullet^*(x, z) = 2$ for all $(x, z)$ for which $1 + z e^{x_4 - x_2} \neq 0$. Applying Singh's algorithm in another way yields

$$(6.3.11) \quad \begin{cases} y_1^{(2)} = e^{x_2 - x_4}((\dot{y}_2 - x_2 - x_4)(x_5 - u_2) + (y_2^{(2)} - x_5 - u_2)) \\ \dot{y}_2 = x_2 + x_4 + e^{x_4} u_1 \end{cases}$$

So $\gamma_1 = \delta_1 = 2$, $\gamma_2 = 1$, $\delta_2 = 2$. Setting $\dot{y}_2 = z$, $y_2^{(2)} = v_2$, $y_1^{(2)} = v_1$ yields the following regular dynamic state feedback solving the LDDDP for (6.3.5) locally around any point $(x_0, z_0)$ with $1 + z_0 - x_{20} - x_{40} \neq 0$:

$$(6.3.12) \quad \begin{cases} \dot{z} = v_2 \\ u_1 = e^{-x_4}(z - x_2 - x_4) \\ u_2 = (1 + z - x_2 - x_4)^{-1}((z - x_2 - x_4)x_5 - x_5 + v_2 - e^{x_4 - x_2} v_1) \end{cases}$$

In this case the largest locally controlled invariant distribution $\hat{\Delta}_\bullet^*$ in ker $dh$ for the system (6.3.5,12) is given by

$$(6.3.13) \quad \hat{\Delta}_\bullet^* = \mathrm{sp}\{\frac{\partial}{\partial x_5}, (z - x_2 - x_4 + 1)\frac{\partial}{\partial x_2} + (z - x_2 - x_4 - 1)\frac{\partial}{\partial x_4}\}$$

Note that the extended disturbance vector field $p_e := (p^T, 0^T)^T = \frac{\partial}{\partial x_5}$ is contained in both $\Delta_\bullet^*$ and $\hat{\Delta}_\bullet^*$. □

Next, we give intrinsic geometric conditions for solvability of the LDDDP using Theorem 6.3.1. Suppose again that $x_0$ is a strongly regular point for Singh's algorithm and that the SLDIODP for (6.3.1) is solvable. Any compensator (6.2.15) obtained from Singh's algorithm fulfills $y_i^{(\epsilon_i)} - v_i$, $i - 1, \ldots, m$, where the constants $\epsilon_i$, $i - 1, \ldots, m$ denote the essential orders (see [GM], [XG]). Hence the decoupling matrix $A(x,z)$ for the system (6.3.1,2.15) is just the identity matrix and the largest locally controlled invariant distribution in ker $dh$ for the compensated system is given by

$$\Delta_e^* - \bigcap_{i=1}^{m} \bigcap_{k=0}^{\epsilon_i-1} \ker dy_i^{(k)}$$ (cf. (2.4.5)). Obviously, $\Delta_e^*$ depends on the partic-

ular application of Singh's algorithm (see also Example 6.3.3), so $\Delta_e^*$ is by no means uniquely defined. However, if the LDDDP for (6.3.1) is solvable, it can be solved by applying any Singh compensator (6.2.15), so the extended disturbance vector fields $p_{ie} := \left(p_i^T \ 0^T\right)^T$, $i - 1, \ldots, r$ are contained in $\Delta_e^*$ for any compensator (6.2.15). Write $P_e := \mathrm{sp}\{p_{1e}, \ldots, p_{re}\}$. Let $0_z$ denote the zero vector in the tangent space $T_z\mathbb{R}^\mu$. Then we have $P_e(x) \subset T_x\mathbb{R}^n \times 0_z \subset T_x\mathbb{R}^n \times T_z\mathbb{R}^\mu$ for all $(x,z)$ in a neighborhood $\mathcal{O}(x_0,z_0)$ of $(x_0,z_0)$. By construction, $\Delta_e^*(x,z) \subset T_x\mathbb{R}^n \times 0_z$ for all $(x,z) \in \mathcal{O}(x_0,z_0)$, because the Singh compensator (6.2.15) is chosen in such a way that $y_i^{(k)} - z_{i,k-\gamma_i+1}$, $i - 1, \ldots, m$, $k - \gamma_i, \ldots, \delta_i-1$, with $\delta_i - \epsilon_i$, $i - 1, \ldots, m$. Hence, $\Delta_e^*$ is certainly contained in $\bigcap_{i=1}^{m} \bigcap_{i=0}^{\delta_i-\gamma_i-1} \ker dz_i^{(k)}$. Note that the vector fields in $\Delta_e^*$ may depend on $z$, while the vector fields in $P_e$ only depend on $x$. Hence $P_e$ is contained in the maximal subdistribution of $\Delta_e^*$ that contains the vector fields in $\Delta_e^*$ that depend essentially on $x$ only. This distribution, which is not necessarily locally controlled invariant (w.r.t. the extended system), can be found by means of the following algorithm.

**Algorithm 6.3.4**

1.  $\Delta_0 := \Delta_e^*$

2.  $\Delta_k := \{\tau \in \Delta_{k-1} | \ [\tau, \frac{\partial}{\partial z_i}] \in \Delta_{k-1} + \mathrm{sp}\{\frac{\partial}{\partial z_i}| \ i - 1, \ldots, \mu\}, \ i - 1, \ldots, \mu\}$

Suppose that the distributions $\Delta_k$ in Algorithm 6.3.4 have constant dimension for all $k$. Then the algorithm converges in a finite number of steps to $\Delta_{k^*}$, and $\Delta_{k^*}$ fulfills the condition

(6.3.14)  $[\Delta_{k^*}, \frac{\partial}{\partial z_i}] \subset \Delta_{k^*} + \mathrm{sp}\{\frac{\partial}{\partial z_i}| \ i - 1, \ldots, \mu\}, \ i - 1, \ldots, \mu$

As can be seen in [vdS1] it follows from (6.3.14) that there exists a basis for $\Delta_{k*}$ of vector fields for which the first $n$ components depend on $x$ only. Since the last $\mu$ components of any vector field in $\Delta_{k*}$ equal zero (recall that $\Delta_e^*(x,z) \subset T_x\mathbb{R}^n \times 0_z$), this implies that the vector fields in $\Delta_{k*}$ depend essentially on $x$ only. It can easily be seen that $\Delta_{k*}$ is the largest subdistribution of $\Delta_e^*$ having this property and that $\Delta_{k*}$ is involutive. Hence $\Delta_{k*}$ is the maximal subdistribution of $\Delta_e^*$ that contains the vector fields in $\Delta_e^*$ that depend essentially on $x$ only. In the sequel this distribution is denoted as $\tilde{\Delta}_e^*$.

**Lemma 6.3.5** The distribution $\tilde{\Delta}_e^*$ obtained by Algorithm 6.3.4 is independent of the way Singh's algorithm is applied.

**Proof** Assume we have applied Singh's algorithm in two ways, yielding $\Delta_{e1}^*$ and $\Delta_{e2}^*$. Assume, furthermore, that by applying Algorithm 6.3.4 to these distributions we obtain $\tilde{\Delta}_{e1}^*$ and $\tilde{\Delta}_{e2}^*$ with $\tilde{\Delta}_{e1}^* \neq \tilde{\Delta}_{e2}^*$. This implies that there exist disturbance vector fields for which the LDDDP is solvable by applying Singh's algorithm in one way, but not in the other way. This contradicts Theorem 6.3.1. Hence $\tilde{\Delta}_{e1}^*$ equals $\tilde{\Delta}_{e2}^*$.                                    □

Let $\tilde{\Delta}^*$ denote the distribution obtained by the projection of $\tilde{\Delta}_e^*(x)$ on $T_x\mathbb{R}^n$ for all $x$. By (6.3.14) this distribution $\tilde{\Delta}^*$ is well-defined (see [vdS1]). By definition $\tilde{\Delta}^*$ contains exactly those vector fields that can be decoupled from the outputs by applying regular dynamic state feedback. These considerations prove the following result.

**Theorem 6.3.6** Consider the square analytic system (6.3.1) and a point $x_0 \in \mathbb{R}^n$. Assume that $x_0$ is a strongly regular point for Singh's algorithm and that the Strong Local Dynamic Input-Output Decoupling Problem is solvable for (6.3.1). Then the Local Dynamic Disturbance Decoupling Problem is solvable if and only if $P \subset \tilde{\Delta}^*$.

**Remark 6.3.7** Obviously, the distribution $\tilde{\Delta}^*$ contains $\Delta^*$, the largest locally controlled invariant distribution in ker $dh$, because $\Delta^*$ exactly contains those disturbance vector fields for which the LDDP is solvable. It follows from comparison of the version of Singh's algorithm in [IM] to calculate a reduced inverse with Algorithm 6.2.1 (see also [dBG]) that the dimension of $\Delta_e^*$ equals the dimension $d$ of the zero dynamics manifold of the system (6.3.1) (which is in general larger than the dimension of $\Delta^*$, cf.

Section 2.5). Hence, if the dimension of $\Delta^*$ equals $d$, then the LDDDP is solvable if and only if the LDDP is.                                                                         □

**Remark 6.3.8** Theorem 3.7 in [IKGM1] gives necessary conditions for disturbance decoupling by dynamic output feedback. If we specialize the results of Isidori et al. to dynamic state feedback, then Theorem 3.7 states the following: "If disturbance decoupling for the system (6.3.1) can be achieved by applying dynamic state feedback with full control (i.e. with $\delta(x,z)$ invertible for all $(x,z)$), then $P \subset \Delta^*$ (i.e. the LDDP is solvable for (6.3.1))". We have proved (see Example 6.1.1) that an analogous result does not hold true if more general compensators of the form (6.3.2) are taken into consideration.                                                                         □

In the foregoing we showed that for square systems for which the SLDIODP is solvable the set of disturbance vector fields that can be decoupled from the outputs can be enlarged by applying a regular dynamic state feedback following from Singh's algorithm. The following example shows that this does not hold true if an arbitrary regular dynamic state feedback solving the SLDIODP is applied.

**Example 6.3.9** Consider the system

$$(6.3.15) \quad \begin{cases} \dot{x}_1 = u_1, \ \dot{x}_2 = x_5, \ \dot{x}_3 = x_4 u_1, \ \dot{x}_4 = u_2, \ \dot{x}_5 = x_1 u_1 \\ y_1 = x_1, \ y_2 = x_3 \end{cases}$$

Note that for (6.3.15) strong input–output decoupling may be obtained by dynamic state feedback, but not by static state feedback. Application of Algorithm 2.3.9 yields $\Delta^* = \text{sp}\{\frac{\partial}{\partial x_2}, \frac{\partial}{\partial x_5}\}$. If the compensator

$$(6.3.16) \quad \dot{z}_1 = v_1, \ u_1 = x_2 z_1, \ u_2 = v_2$$

is applied to (6.3.15), then we have for the system (6.3.15,16) $\Delta_e^* = \text{sp}\{\frac{\partial}{\partial x_5}, -x_2\frac{\partial}{\partial x_2} + z_1\frac{\partial}{\partial z_1}\}$. Note that $\frac{\partial}{\partial x_2} \notin \Delta_e^*$. This implies that while the outputs are independent of disturbances that enter the system (6.3.15) via the $x_2$-equation, this independence is destroyed if the dynamic state feedback (6.3.16) which is regular around any point $(x_0, z_0)$ with $x_{20} \neq 0$ is applied to (6.3.15).                                                                         □

## 6.4 The Local Dynamic Disturbance Decoupling Problem with Stability

In this section the LDDDPS is solved hereby extending the results in Chapter 3. Consider the square analytic system

$$(6.4.1) \quad \begin{cases} \dot{x} = f(x) + g(x)u + p(x)q, & f(0) = 0, \ x \in \mathbb{R}^n, \ u \in \mathbb{R}^m, \ q \in \mathbb{R}^r \\ y = h(x), & h(0) = 0, \ y \in \mathbb{R}^m \end{cases}$$

Assume that

(E1) The Strong Local Dynamic Input–Output Decoupling Problem for (6.4.1) is solvable;

(E2) $x = 0$ is a strongly regular point for Singh's algorithm.

In the sequel we suppose that the conditions (6.2.9) and (6.2.10) hold with $\tilde{y}_{i0}^{(j)} = 0$ ($1 \le i \le n-1$, $i \le j \le n-1$). Let

$$(6.4.2) \quad \begin{cases} \dot{z} = Az + Bv, & x \in \mathbb{R}^n, \ z \in \mathbb{R}^\mu \\ u = \gamma(x,z) + \delta(x,z)v, & \gamma(0,0) = 0, \ u \in \mathbb{R}^m, \ v \in \mathbb{R}^m \end{cases}$$

be an arbitrary Singh compensator locally defined on a neighborhood $\mathcal{O}$ of $(x,z) = (0,0)$, i.e. a compensator that solves the SLDIODP for (6.4.1), cf. Section 6.2, equations (6.2.15,16). The system (6.4.1,2) is described by the following equations

$$(6.4.3) \quad \begin{cases} \dot{x} = f(x) + g(x)\gamma(x,z) + g(x)\delta(x,z)v + p(x)q \\ \dot{z} = Az + Bv \\ y = h(x) \end{cases}$$

written in short-hand notation as

$$(6.4.4) \quad \begin{cases} \dot{x}_e = f_e(x_e) + g_e(x_e)v + p_e(x)q \\ y = h_e(x_e) = h(x) \end{cases}$$

where $x_e = \binom{x}{z}$. In the sequel the subindex $e$ refers to the extended system (6.4.3) (or (6.4.4)). In order to see if the LDDDPS for (6.4.1) is solvable we try to find the maximal stabilizability distribution $(\Delta_e^*)_s$ in the kernel of the output mapping of the system (6.4.3) using the theory from Chapter

3. The LDDDPS for (6.4.1) is solvable then if $P_e \subset (\Delta_e^*)_s$. The following example shows that there exist systems for which disturbance decoupling with stability can be obtained by dynamic feedback, while this is not possible by applying static feedback.

**Example 6.4.1**  Consider the system

$$(6.4.5) \quad \begin{cases} \dot{x}_1 = -x_1+(1+x_2)u_1, \quad \dot{x}_2 = x_5, \quad \dot{x}_3 = x_4+u_1, \quad \dot{x}_4 = x_2+u_2 \\ \dot{x}_5 = -x_3+x_4-2x_5+u_1+u_2+q \\ y_1 = x_1, \quad y_2 = x_3 \end{cases}$$

It can easily be checked that $\Delta^* = 0$. Note that with

$$(6.4.6) \quad x_4+u_1 = z, \quad \dot{z} = v_2$$

we have

$$(6.4.7) \quad \begin{cases} \dot{y}_2 = x_4+u_1 = z, \quad y_2^{(2)} = v_2 \\ \dot{y}_1 = -x_1+(1+x_2)u_1 = -x_1+(1+x_2)(z-x_4) \\ y_1^{(2)} = x_1-(1+x_2)(z-x_4)+x_5(z-x_4)+(1+x_2)(v_2-x_2-u_2) =: v_1 \end{cases}$$

Hence the regular dynamic state feedback

$$(6.4.8) \quad \begin{cases} \dot{z} = v_2 \\ u_1 = z-x_4 \\ u_2 = \dfrac{1}{1+x_2}(x_1-(1+x_2)(z-x_4)+x_5(z-x_4)-x_2(1+x_2))-\dfrac{1}{1+x_2}v_1+v_2 \end{cases}$$

solves the LDDDP for (6.4.5) and $\Delta_e^*$ is given by

$$(6.4.9) \quad \Delta_e^* = sp\{\dfrac{\partial}{\partial x_5}, \quad (1+x_2)\dfrac{\partial}{\partial x_2} + (z-x_4)\dfrac{\partial}{\partial x_4}\}$$

The zero dynamics manifold of the system (6.4.5,8) equals the leaf $L_e$ of $\Delta_e^*$ through $(x,z) = (0,0)$, so $L_e = \{(x,z)| x_1 = x_3 = x_4 = z = 0\}$ and the zero dynamics are given by

$$(6.4.10) \quad \dot{x}_2 = x_5, \quad \dot{x}_5 = -x_2-2x_5$$

Note that these dynamics are exponentially stable, hence the stable invariant manifold $S_e$ through $(x,z) = (0,0)$ is exactly the leaf $L_e$.

Consequently, for the system (6.4.5,8) the maximal stabilizability distribution $(\Delta_e^*)_s$ is equal to $\Delta_e^*$. The linearization of the dynamics (6.4.5,8) modulo $(\Delta_e^*)_s$ is also stabilizable. So the LDDDPS for (6.4.5) can be solved by applying the compensator (6.4.8) together with the stabilizing state feedback

(6.4.11) $\quad v_1 = x_1 - 4z + w_1, \quad v_2 = x_3 + 2(1+x_2)(z-x_4) + w_2$ $\qquad\qquad$ $\square$

It follows from the definition of the compensator (6.2.15) (or (6.4.2)) that if $x = 0$ is a strongly regular point for Singh's algorithm then the zero dynamics manifold $L_e$ and the zero dynamics are the same for any application of Singh's algorithm. Moreover, the equations that describe the zero dynamics for the compensated system are the same as the equations that give the zero dynamics of the original system (6.4.1) together with the equations $z = 0$. (Obviously, the restricted zero dynamics and the zero dynamics coincide, because the compensated system is square and strongly input—output decoupled, cf. Section 2.5.) This implies that the stable invariant manifold $S_e$ through $(x,z) = (0,0)$ and the dynamics restricted to $S_e$ are uniquely defined. If $S_e$ and $L_e$ coincide (i.e. if the zero dynamics have an exponentially stable equilibrium), then the LDDDPS for (6.4.1) is solvable by any application of a compensator of the form (6.4.2) followed by a feedback that stabilizes the linearization of the system (6.4.1,2) modulo $\Delta_e^*$. (Note that these linearized dynamics are controllable, because (6.4.2) is a decoupling feedback (cf. Section 2.5).) This proves the following theorem.

**Theorem 6.4.2** Consider the square analytic system (6.4.1). Assume that (E1) and (E2) hold and that $x = 0$ is an exponentially stable equilibrium of the zero dynamics of the system (6.4.1). Then the Local Dynamic Disturbance Decoupling Problem with Stability for (6.4.1) is solvable if and only if $P_e \subset \Delta_e^*$.

This result states that for square strongly dynamically input—output decouplable systems for which the zero dynamics have an exponentially stable equilibrium the Local Dynamic Disturbance Decoupling Problem with Stability is solvable if and only if the Local Dynamic Disturbance Decoupling Problem is. In the sequel we call such a system *dynamically exponentially minimum phase*.

**Remark 6.4.3** In Section 2.5 the notion of exponentially minimum phase system is introduced for square strongly input–output decouplable systems only. The term dynamically exponentially minimum phase refers to the fact that the extended system is strongly input–output decouplable. Note that in [BI3] a system is called minimum phase if the zero dynamics have an asymptotically (but not necessarily exponentially) stable equilibrium (say $x_0$). Byrnes and Isidori prove that a minimum phase system fulfilling some invertibility condition can be asymptotically stabilized around $x_0$ by smooth static state feedback. They prove this by redesigning the outputs in such a way that the system with these new outputs has the property that all relative degrees are equal to one and that the decoupling matrix has full row rank. It follows from Theorem 6.4.2 that for dynamically exponentially minimum phase systems an analogous result can be obtained easier, at least if dynamic state feedback is allowed.                                  □

In case $S_e$ does not coincide with $L_e$ (and so $(\Delta_e^*)_s \neq \Delta_e^*$) things are more complicated. We show in the sequel that if the LDDDPS is solvable for (6.4.1) by applying a compensator (6.4.2) together with a stabilizing feedback, then the LDDDPS is solvable independent of the choice of the $\tilde{y}_i^{(j)}$ in Singh's algorithm. Assume that

(E3)  The linearization of the system (6.4.1) around $x = 0$ is stabilizable;

Suppose that $(\Delta_e^*)_s$ for the system (6.4.1,2) exists. (This is the case if e.g. the conditions in Section 3.2 hold for the extended system (6.4.1,2).) Suppose, moreover, that

$$(6.4.12) \quad \begin{cases} \dot{z}^1 = A^1 z^1 + B^1 v, & x \in \mathbb{R}^n,\ z^1 \in \mathbb{R}^\mu \\ u = \gamma^1(x,z^1) + \delta^1(x,z^1)v,\ \gamma^1(0,0) = 0\ ,\ u \in \mathbb{R}^m,\ v \in \mathbb{R}^m \end{cases}$$

is another Singh compensator. (In the sequel the superscript $^1$ also refers to the system (6.4.1,12).) It follows from Proposition 6.2.5 that there exists an analytic coordinate transformation $(x,z^1) = \Phi(x,z)$ transforming (6.4.2) in (6.4.12). This immediately implies that $\frac{\partial\Phi}{\partial x_e}(\Delta_e^*) = (\Delta_e^*)^1$.

Let $\Delta := \frac{\partial\Phi}{\partial x_e}((\Delta_e^*)_s)$. Then $\Delta$ is contained in $(\Delta_e^*)^1$, invariant under $f_e^1$ and $g_e^1$ and the leaf of $\Delta$ through $(x,z^1) = (0,0)$ is the stable invariant manifold $S_e^1 = \Phi(S_e)$. Hence, $\Delta$ equals the maximal stabilizability distribution $(\Delta_e^*)_s^1$ in $(\Delta_e^*)^1$. This result shows that existence of $(\Delta_e^*)_s$ does not depend on the particular application of Singh's algorithm.

The main result (Theorem 6.4.5) follows straightforward from these considerations and Lemma 6.4.4.

**Lemma 6.4.4** Consider the square analytic system (6.4.1). Assume that (E1), (E2) and (E3) hold. Then the linearization of the system (6.4.1,2) around the equilibrium $(x,z) = (0,0)$ is stabilizable.

The proof is omitted. It follows from carefully checking Singh's algorithm.

**Theorem 6.4.5** Consider the square analytic system (6.4.1). Assume that the conditions (E1), (E2) and (E3) hold. Let (6.4.2) be an arbitrary analytic Singh compensator and suppose that $(\Delta_e^*)_s$ exists for the system (6.4.1,2). Then the Local Dynamic Disturbance Decoupling Problem with Stability for (6.4.1) is solvable if $P_e \subset (\Delta_e^*)_s$.

We conclude with an example where $(\Delta_e^*)_s \neq \Delta_e^*$.

**Example 6.4.6** Consider the stabilizable system

$$(6.4.13) \quad \begin{cases} \dot{x}_1 = x_2 + u_1, & \dot{x}_2 = -x_5 + (1+x_4)u_1, & \dot{x}_3 = x_1 - x_3 \\ \dot{x}_4 = x_2 + 2x_4 + u_1, & \dot{x}_5 = x_1 + x_3 u_1 + u_2 \\ y_1 = x_1, & y_2 = x_2 \end{cases}$$

For this system $\Delta^* = 0$. The regular dynamic state feedback

$$(6.4.14) \quad \begin{cases} \dot{z} = v_1 \\ u_1 = z - x_2 \\ u_2 = -x_1 - x_3(z-x_2) + (z+2x_4)(z-x_2) + (1+x_4)\big[x_5 - (1+x_4)(z-x_2)\big] \\ \qquad\qquad + (1+x_4)v_1 - v_2 \end{cases}$$

solves the SLDIODP for (6.4.13).

Now $\Delta_e^* = \text{sp}\{\frac{\partial}{\partial x_3}, \frac{\partial}{\partial x_4} + (z-x_2)\frac{\partial}{\partial x_5}\}$. The zero dynamics manifold is given by $\{(x,z) \mid x_1 = x_2 = x_5 = z = 0\}$ and the zero dynamics by

$$(6.4.15) \quad \dot{x}_3 = -x_3, \quad \dot{x}_4 = 2x_4$$

Hence, the stable invariant manifold $S_e$ equals the $x_3$-axis in the $(x,z)$-plane and $(\Delta_e^*)_s = \text{sp}\{\frac{\partial}{\partial x_3}\}$, as can easily be seen from the equations (6.4.13,14):

$$(6.4.16) \quad \begin{cases} \dot{x}_1 = z, & \dot{x}_2 = -x_5 + (1+x_4)(z-x_2) \\[2mm] \dot{x}_3 = x_1 - x_3, & \dot{x}_4 = 2x_4 + z \\[2mm] \dot{x}_5 = (z+2x_4)(z-x_2) + (1+x_4)\big(x_5 - (1+x_4)(z-x_2)\big) + (1+x_4)v_1 - v_2 \\[2mm] \dot{z} = v_1 \\[2mm] y_1 = x_1, & y_2 = x_2 \end{cases}$$

The feedback

$$(6.4.17) \quad v_1 = \tfrac{1}{2}x_1 - 13\tfrac{1}{2}x_4 - 5z + w_1, \quad v_2 = x_2 + 2x_5 + w_2$$

stabilizes the dynamics of the system (6.4.16) modulo $(\Delta_e^*)_s$. Hence, the regular dynamic state feedback (6.4.14,17) solves the LDDDPS for (6.4.13).

□

## 7.  CONCLUSIONS

Several design problems in linear systems theory have been treated fruitfully by the geometric approach. Motivated by the success of the geometric
theory, researchers in nonlinear systems theory tried to translate several
geometric concepts to a nonlinear context using differential geometric
tools. This led for instance to the definition of (local) controlled invariance and the solution of a local version of the Disturbance Decoupling
Problem. So far, only a few papers have been published on the solution of
design problems where the extra requirement of stability is imposed on the
resulting feedback system, although the general problem of feedback stabilization is a very active research area. These considerations motivated the
problem that is addressed in this monograph.

In the main part of this monograph the Local Disturbance Decoupling Problem
with Stability for nonlinear systems is considered. This problem consists
in finding conditions under which there exists a locally defined regular
static state feedback that decouples the outputs from the disturbances and
exponentially stabilizes the equilibrium of the modified drift dynamics of
the feedback system. For systems for which the linearization of the
dynamics around an equilibrium is stabilizable two methods are proposed to
solve this problem.

In the first method stabilizability distributions for nonlinear systems are
introduced and it is shown that under certain regularity assumptions the
maximal stabilizability distribution $\Delta_s^*$ in the kernel of the output mapping
exists and that the Local Disturbance Decoupling Problem with Stability is
solvable if and only if the disturbance vector fields are contained in $\Delta_s^*$.
This distribution forms a nonlinear analogue of the maximal stabilizability
subspace $V_s^*$ for linear systems. One may try to find this distribution by
prolongation of the stable invariant submanifold $S_0$ of the integral manifold $M_0$ through the equilibrium of the largest locally controlled invariant
distribution $\Delta^*$ in the kernel of the output mapping, that is, $S_0$ is the
submanifold of $M_0$ that is invariant under the drift dynamics of the system
restricted to $M_0$ and tangent to the stable subspace of the linearization of
these dynamics at the equilibrium. It is shown (by means of an example)
that this method does not always generate a locally controlled invariant

distribution. If this happens, then the dimension of the distribution $\Delta_s^*$ is strictly less than the dimension of the stable invariant manifold $S_0$. This is a nonlinear phenomenon, because for linear systems these dimensions are always equal.

A second more "problem oriented" method searches for the smallest locally controlled invariant distribution $(\Delta^P)_*$ in the kernel of the output mapping containing the disturbance vector fields as well as the largest local controllability distribution in $\Delta^*$. If the linearization of the dynamics of the system restricted to the leaf of $(\Delta^P)_*$ through the equilibrium is stabilizable, then the Local Disturbance Decoupling Problem with Stability is solvable.

Since stabilizability of the linearization of the nonlinear system around an equilibrium is a necessary condition for solvability of the Local Disturbance Decoupling Problem with Stability (LDDPS), one may wonder if solvability of the Disturbance Decoupling Problem with Stability (DDPS) for the linearization is sufficient for solvability of the LDDPS for the nonlinear system. The answer is negative as can easily be seen from an example. The main result in Chapter 5 states that if the Local Disturbance Decoupling Problem for a strongly input–output decouplable nonlinear system is solvable, then the Local Disturbance Decoupling Problem with Stability is solvable if and only if $(\Delta^P)_*(0)$, regarded as a subspace of $\mathbb{R}^n$, is contained in $\mathcal{V}_s^*$. Moreover, a linear feedback that solves the DDPS for the linearization of the nonlinear system around the equilibrium is the linearization of a feedback that solves the LDDPS for the nonlinear system by making $(\Delta^P)_*$ invariant if and only if the linear feedback makes $(\Delta^P)_*(0)$ invariant.

Another typical nonlinear phenomenon is treated in the last part of this monograph. For linear systems it is well-known that if the Disturbance Decoupling Problem (with Stability) is solvable by applying dynamic state feedback it can also be solved by applying static state feedback. It is shown by means of an example that it is possible that disturbance decoupling can be obtained for a nonlinear system if an integrator is added to one of the inputs first, while disturbance decoupling cannot be achieved by applying static state feedback. This example gives rise to the definition of the Local Dynamic Disturbance Decoupling Problem and a version of this problem including a stability requirement.

For square systems that are strongly dynamically input–output decouplable these problems may be tackled by applying Singh's algorithm. The main result on the Local Dynamic Disturbance Decoupling Problem states that, if this problem is solvable, then it can be solved using Singh's algorithm. Furthermore, solvability of the Local Dynamic Disturbance Decoupling Problem with Stability can be checked as follows: first, apply a compensator obtained by an arbitrary application of Singh's algorithm to the given system and then use the techniques developed in Chapter 3. (Note that the system and compensator together form a strongly input–output decoupled system.) It is shown that solvability of the Local Dynamic Disturbance Decoupling Problem (with Stability) does not depend on the specific application of the algorithm. Finally, for a dynamically exponentially minimum phase system (that is, a square strongly dynamically input–output decouplable system for which the zero dynamics are exponentially stable around the equilibrium) the Local Dynamic Disturbance Decoupling Problem with Stability is solvable if and only if the Local Dynamic Disturbance Decoupling Problem is.

This monograph leaves some interesting questions for further research. A few of these open problems are listed below.

(i)    Does there exist an algorithm to calculate $\Delta_s^*$ if prolongation of the stable invariant manifold does not generate $\Delta_s^*$? Is there a method to circumvent the prolongation, which from a computational point of view is inattractive?

(ii)   Is it possible to extend the results obtained in Section 6.4 to nonsquare and/or noninvertible nonlinear systems (cf. [HNvdW2] where the results of Section 6.3 are generalized)? If so, is the set of systems for which the Local Dynamic Disturbance Decoupling Problem with Stability is solvable larger than the set for which the Local Disturbance Decoupling Problem with Stability is solvable?

(iii)  Is it possible to generalize the results in this monograph to the nonregular Disturbance Decoupling Problem with Stability, i.e. the problem of finding conditions under which disturbance decoupling and exponential stability can be obtained in case *singular* feedbacks are allowed for. An extension of this kind might be useful in order to find a solution for the Nonlinear Model Matching Problem with Stability. It is proved by Huijberts (see [Hu] and references therein) that the Nonlinear Model Matching Problem is solvable if and only if the nonregular Dynamic Disturbance Decoupling Problem with disturbance measurements is.

(iv)   To what extend can the *local* results on disturbance decoupling with stability given here be globalized? Recently, some articles have been published on (semi)global stabilization of nonlinear control systems (see e.g. [Su]) which show some limitations to this globalization.

Finally, systems theory is a research area originating from measurement and control methods used in engineering practice to steer "real-world" processes. So in order that the mathematical theory developed in this monograph becomes meaningful for practical problems, it is needed to find (simplified) models of existing processes to which the theory is applicable. However, much work still needs to be done before the theory can be applied in practice.

# REFERENCES

[Ae1]   D. Aeyels, "Stabilization of a class of nonlinear systems by a smooth feedback control", Syst. Contr. Lett., **5**, pp. 289-294, 1985.

[Ae2]   D. Aeyels, "Local and global stabilizability for nonlinear systems", in **Theory and Applications of Nonlinear Control Systems** (eds. C.I. Byrnes, A. Lindquist), North-Holland, Amsterdam, pp. 93-105, 1986.

[AG]    M.A. Aiserman, F.R. Gantmacher, **Die Absolute Stabilität von Regelsystemen**, Oldenburg, München, 1965.

[Ak]    O. Akhrif, **Nonlinear Adaptive Control with Applications to Flexible Structures**, Ph.D. Thesis, Univ. of Maryland, 1989.

[Ba]    A. Bacciotti, "The local stabilizability problem for nonlinear systems", IMA Journal of Math. Contr. & Inf., **5**, pp. 27-39, 1988.

[dBG]   M.D. di Benedetto, J.W. Grizzle, "Intrinsic notions of regularity for local inversion, output nulling and dynamic extension of nonsquare systems", C-TAT, **6**, pp. 357-381, 1990.

[dBGM]  M.D. di Benedetto, J.W. Grizzle, C.H. Moog, "Rank invariants of nonlinear systems", SIAM J. Contr. Opt., **27**, pp. 658-672, 1989.

[Bh1]   S.P. Bhattacharyya, "Disturbance rejection in linear systems", Int. J. Systems Sci., **5**, pp. 633-637, 1974.

[Bh2]   S.P. Bhattacharyya, "Frequency domain conditions for disturbance rejection", IEEE Trans. Aut. Contr., **AC-25**, pp. 1211-1213, 1980.

[BI1]   C.I. Byrnes, A. Isidori, "A frequency domain philosophy for nonlinear systems with applications to stabilization and to adaptive control", Proc. 23rd IEEE Conf. on Dec. and Contr., Las Vegas, pp. 1569-1573, 1984.

[BI2]   C.I. Byrnes, A. Isidori, "Nonlinear disturbance decoupling with stability", Proc. 26th IEEE Conf. on Dec. and Contr., Los Angeles, pp. 513-518, 1987.

[BI3]   C.I. Byrnes, A. Isidori, "Local stabilization of minimum-phase nonlinear systems", Syst. Contr. Lett., **11**, pp. 9-17, 1988.

[BM1]   G. Basile, G. Marro, "Controlled and conditioned invariant subspaces in linear system theory", J. Optim. Theory Appl., **3**, pp. 306-315, 1969.

[BM2]   G. Basile, G. Marro, "Dual-lattice theorems in the geometric approach", J. Optim. Theory Appl., **48**, pp. 229-244, 1986.

[BMP]   G. Basile, G. Marro, A. Piazzi, "A new solution to the disturbance localization problem with stability and its dual", Proc. Int. AMSE Conf. on Modelling and Simulation, Athens, pp. 19-27, 1984.

[Bo]     W.M. Boothby, **An Introduction to Differential Manifolds and Riemannian Geometry**, Academic Press, New York, 1975.

[Br]     R.W. Brockett, "Feedback invariants for nonlinear systems", Proc. 7th IFAC World Congress Helsinki, pp. 1115–1120, 1978.

[Ca]     J. Carr, **Applications of Centre Manifold Theory**, Springer, New York, 1981.

[Cr]     J.J. Craig, **Introduction to Robotics: mechanics and control**, Addison–Wesley, Reading, 1986.

[DM]     J. Descusse, C.H. Moog, "Decoupling with dynamic compensation for strong invertible affine nonlinear systems", Int. J. Contr., **42**, pp. 1387–1398, 1985.

[Fl]     M. Fliess, "A note on the invertibility of nonlinear input–output differential systems", Syst. Contr. Lett., **8**, pp. 147–151, 1986.

[Fr]     E. Freund, "The structure of decoupled nonlinear systems", Int. J. Contr., **21**, pp. 443–450, 1975.

[GI]     J.W. Grizzle, A. Isidori, "Block noninteracting control with stability via static state feedback", Math. Contr. Sign. Syst., **2**, pp. 315–341, 1989.

[GM]     A. Glumineau, C.H. Moog, "The essential orders and the nonlinear decoupling problem", Int. J. Contr., **50**, pp. 1825–1834, 1989.

[GN]     L.C.J.M. Gras, H. Nijmeijer, "Decoupling in nonlinear systems: from linearity to nonlinearity", IEE Proc. Pt.D., **136**, pp. 53–62, 1989.

[Gr]     L.T. Grujič, "On absolute stability and the Aizerman conjecture", Automatica, **17**, pp. 335–349, 1981.

[Ha]     W. Hahn, **Stability of Motion**, Springer, Berlin, 1967.

[Har]    P. Hartman, **Ordinary Differential Equations**, Wiley, New York, 1964.

[Hau]    M.L.J. Hautus, "(A,B)-invariant and stabilizability subspaces, a frequency domain description", Automatica, **16**, pp. 703–707, 1980.

[Hi1]    R.M. Hirschorn, "Invertibility of multivariable nonlinear control systems", IEEE Trans. Aut. Contr., **AC–24**, pp. 855–865, 1979.

[Hi2]    R.M. Hirschorn, "(A,B)-invariant distributions and disturbance decoupling of nonlinear systems", SIAM J. Contr. Opt., **19**, pp. 1–19, 1981.

[HN]     H.J.C. Huijberts, H. Nijmeijer, "Local nonlinear model matching: from linearity to nonlinearity", Automatica, **26**, pp. 173–183, 1990.

[HNvdW1]  H.J.C. Huijberts,  H. Nijmeijer,  L.L.M. van der Wegen,  "Dynamic disturbance decoupling for nonlinear systems", to appear in SIAM J. Contr. Opt.

[HNvdW2]  H.J.C. Huijberts,  H. Nijmeijer,  L.L.M. van der Wegen,  "Dynamic disturbance decoupling for nonlinear systems: the nonsquare and noninvertible case", in **Controlled Dynamical Systems** (eds. B. Bonnard, B. Bride, J.P. Gauthier, I. Kupka), Birkhäuser, Boston, 1991.

[HNvdW3]  H.J.C. Huijberts, H. Nijmeijer, L.L.M. van der Wegen, "Nonlinear dynamic disturbance decoupling", to appear in Proceedings of the CDC 1990.

[HNvdW4]  H.J.C. Huijberts, H. Nijmeijer, L.L.M. van der Wegen, "Minimality of dynamic input–output decoupling for nonlinear systems", submitted.

[HSM]  J. Hauser, S. Sastry, G. Meyer, "Nonlinear controller design for flight control systems", preprints of IFAC Symposium on Nonlinear Control Systems Design, Capri, Italy, pp. 136–141, 1989.

[HSuM]  L.R. Hunt,  R. Su,  G. Meyer,  "Design for multi–input nonlinear systems",  in  **Differential  Geometric  Control  Theory**  (eds. R.W. Brockett,  R.S. Millman,  H. Sussmann)  Birkhäuser,  Boston, pp. 268–298, 1983.

[Hu]  H.J.C. Huijberts, **Dynamic Feedback in Nonlinear Synthesis Problems**, Ph.D. Thesis, University of Twente, 1991.

[IG]  A. Isidori, J.W. Grizzle, "Fixed modes and nonlinear noninteracting control with stability", IEEE Trans. Aut. Contr., **AC–33**, pp. 907–914, 1988.

[IKGM1]  A. Isidori,  A.J. Krener,  C. Gori–Giorgi,  S. Monaco,  "Nonlinear decoupling via feedback: a differential geometric approach", IEEE Trans. Aut. Contr., **AC–26**, pp. 331–345, 1981.

[IKGM2]  A. Isidori,  A.J. Krener,  C. Gori–Giorgi,  S. Monaco,  "Locally (f,g)–invariant distributions", Syst. Contr. Lett., **1**, pp. 12–15, 1981.

[IM]  A. Isidori, C.H. Moog, "On the nonlinear equivalent of the notion of transmission zeros", in **Modelling and Adaptive Control** (eds. C.I. Byrnes, A. Kurzhanski), Lect. Notes Contr. Inf. Sci., **105**, Springer, Berlin, pp. 445–471, 1988.

[Is]  A. Isidori, **Nonlinear Control Systems** (2nd edition), Springer, Berlin, 1989.

[JQ]  V. Jurdjevic, J.P. Quinn, "Controllability and Stability", J. Diff. Eqs., **28**, pp. 381–389, 1978.

[JR]  B. Jakubczyk, W. Respondek, "On linearization of control systems", Bull. Acad. Polonaise Sci. Ser. Sci. Math., **28**, pp. 517–522, 1980.

[KI]    A.J. Krener, A. Isidori, "Nonlinear zero distributions", Proc. 19th IEEE Conf. on Dec. and Contr., Albuquerque, pp. 665-668, 1980

[Kr1]   A.J. Krener, "A generalization of Chow's theorem and the Bang-Bang theorem to nonlinear control problems", SIAM J. Contr. Opt., 12, pp. 43-52, 1974.

[Kr2]   A.J. Krener, "(Ad f,g), (ad f,g) and locally (ad f,g) invariant and controllability distributions", SIAM J. Contr. Opt., 23, pp. 523-549, 1985.

[MG]    C.H. Moog, A. Glumineau, "Le problème du réjet de perturbations mesurables dans les systèmes non linéaires-application à l'amarrage en un seul point des grands pétroliers", Outils et Modèles Mathématiques pour l'Automatique, l'Analyse de Systèmes et le Traitement du Signal, Vol III (ed. I.D. Landau), Editions du CNRS, Paris, pp. 689-698, 1983.

[Mo]    C.H. Moog, "Nonlinear decoupling and structure at infinity", Math. Contr. Sign. Syst., 1, pp. 257-268, 1988.

[Nij1]  H. Nijmeijer, "Invertibility of affine nonlinear control systems: a geometric approach", Syst. Contr. Lett., 2, pp. 163-168, 1982.

[Nij2]  H. Nijmeijer, "The triangular decoupling problem for nonlinear control systems", Nonl. Anal. Theory, Methods and Appl., 8, pp. 273-279, 1984.

[NR]    H. Nijmeijer, W. Respondek, "Dynamic input-output decoupling of nonlinear control systems", IEEE Trans. Aut. Contr., AC-33, pp. 1065-1070, 1988.

[NvdS1] H. Nijmeijer, A.J. van der Schaft, "Controlled invariance for nonlinear systems", IEEE Trans. Aut. Contr., AC-27, pp. 904-914, 1982.

[NvdS2] H. Nijmeijer, A.J. van der Schaft, "The disturbance decoupling problem for nonlinear control systems", IEEE Trans. Aut. Contr., AC-28, pp. 621-623, 1983.

[NvdS3] H. Nijmeijer, A.J. van der Schaft, "Controlled invariance for nonlinear systems: two worked examples", IEEE Trans. Aut. Contr., AC-29, pp. 361-364, 1984.

[NvdS4] H. Nijmeijer, A.J. van der Schaft, Nonlinear Dynamical Control Systems, Springer, New York, 1990.

[NS]    H. Nijmeijer, J.M. Schumacher, "The regular local noninteracting control problem for nonlinear control systems", SIAM J. Contr. Opt., 24, pp. 1232-1245, 1986.

[Re]    W.T. Reid, Ordinary Differential Equations, Wiley, New York, 1971.

[Res]   W. Respondek, "Nonlinear dynamic disturbance decoupling", in Controlled Dynamical Systems (eds. B. Bonnard, B. Bride, J.P. Gauthier, I. Kupka), Birkhäuser, Boston, 1991.

[RN]        W. Respondek, H. Nijmeijer, "On local right-invertibility of nonlinear control systems", C-TAT, **4**, pp. 325–348, 1988.

[vdS1]      A.J. van der Schaft, "Observability and controllability for smooth nonlinear systems", SIAM J. Contr. Opt., **20**, pp. 338–354, 1982.

[vdS2]      A.J. van der Schaft, "On clamped dynamics of nonlinear systems", in **Analysis and Control of Nonlinear Systems** (eds. C.I. Byrnes, C.F. Martin, R.E. Saeks), Elsevier, Amsterdam, pp. 499–506, 1988.

[Sch]       J.M. Schumacher, "(C,A)-invariant subspaces: some facts and uses", Report 110, VU, 1979.

[Si1]       S.N. Singh, "Decoupling of invertible nonlinear systems with state feedback and precompensation", IEEE Trans. Aut. Contr., **AC–25**, pp. 1237–1239, 1980.

[Si2]       S.N. Singh, "A modified algorithm for invertibility in nonlinear systems", IEEE Trans. Aut. Contr., **AC–26**, pp. 595–598, 1981.

[SJ]        H. Sussmann, V. Jurdjevic, "Controllability of nonlinear systems", J. Diff. Eqs., **12**, pp. 95–116, 1972.

[So]        E.D. Sontag, "Smooth stabilization implies coprime factorization", IEEE Trans. Aut. Contr., **AC–34**, pp. 435–443, 1989.

[Sp]        M. Spivak, **A Comprehensive Introduction to Differential Geometry**, Publish or Perish, Boston, MA, 1970.

[SR]        S.N. Singh, W.J. Rugh, "Decoupling in a class of nonlinear systems by state variable feedback", ASME J. Dyn. Syst. Meas. Contr., **94**, pp. 323–329, 1972.

[Su]        H.J. Sussmann, "Limitations on the stabilizability of globally minimum phase systems", IEEE Trans. Aut. Contr., **AC–35**, pp. 117–118, 1990.

[Tr]        H.L. Trentelman, **Almost Invariant Subspaces and High–Gain Feedback**, CWI Tract 29, Amsterdam, 1986.

[Ts]        J. Tsinias, "Stabilization of nonlinear control systems to subspaces", Int. J. Contr., **46**, pp. 529–535, 1987.

[VV]        M. Vidyasagar, A. Vannelli, "New relationships between input-output and Lyapunov stability", IEEE Trans. Aut. Contr., **AC–27**, pp. 481–483, 1982.

[vdW1]      L.L.M. van der Wegen, "On the use of stable distributions in design problems", preprints of IFAC Symposium on Nonlinear Control Systems Design, Capri, Italy, pp. 325–330, 1989.

[vdW2]      L.L.M. van der Wegen, "Another approach to the local disturbance decoupling problem with stability for nonlinear systems", in **Robust Control of Linear Systems and Nonlinear Control** (eds. M.A. Kaashoek, J.H. van Schuppen, A.C.M. Ran), Birkhäuser, Boston, pp. 465–472, 1990.

[Wi]      J.C. Willems, "Almost invariant subspaces: an approach to high gain feedback design – part 1: Almost controlled invariant subspaces", IEEE Trans. Aut. Contr., **AC–26**, pp. 235–252, 1981.

[vdWN1]   L.L.M. van der Wegen, H. Nijmeijer, "A note on disturbance decoupling with stability for nonlinear systems", in **Analysis and Optimization of Systems** (eds. A. Bensoussan, J.L. Lions), Lect. Notes Contr. Inf. Sci., **111**, Springer, Berlin, pp. 115–126, 1988.

[vdWN2]   L.L.M. van der Wegen, H. Nijmeijer, "The local disturbance decoupling problem with stability for nonlinear systems", Syst. Contr. Lett., **12**, pp. 139–149, 1989.

[Wo]      W.M. Wonham, **Linear Multivariable Control: A Geometric Approach**, Springer, New York, 1979.

[XG]      X.-H. Xia, W.-B. Gao, "A minimal order compensator for decoupling a nonlinear system", preprint, 1989.

In this appendix we prove Lemma 1.16. For convenience we recall this lemma here.

**Lemma 1.16**  Consider the smooth nonlinear system

(1)
$$\begin{cases} \dot{x} = f(x) + g(x)u + p(x)q, & x \in \mathbb{R}^n, \ u \in \mathbb{R}^m, \ q \in \mathbb{R}^r \\ y = h(x), & y \in \mathbb{R}^\ell \end{cases}$$

Assume that $x = 0$ is a locally exponentially stable equilibrium of $f$. Suppose that the vector fields $p_i$, $i = 1,\ldots,r$ are bounded. Then the system (1) is locally BDBS-stable.

Precisely, consider the system (1) with $u = 0$ and the disturbances as inputs. Then there exist neighborhoods $O$ and $\tilde{O}$ of $x = 0$ and a constant $D$ such that if $x_0 \in \tilde{O}$ and $|q(t)| \leq D$ for all $t$ ($|\ |$ denotes the Euclidean norm), then $x(t) \in O$ for all positive $t$.

In the proof we use the following version of Gronwall's lemma.

**Lemma A1**  ([Re])    Let $k \geq 0$ and $T_0 > 0$. Suppose that $\varphi$ and $\psi$ denote continuous functions from $[0,T_0)$ in $\mathbb{R}$ and that

(2)
$$\varphi(t) \leq \psi(t) + k \int_0^t \varphi(s)ds, \quad 0 \leq t < T_0$$

Then

(3)
$$\varphi(t) \leq \psi(t) + k \int_0^t \exp[k(t-s)]\psi(s)ds, \quad 0 \leq t < T_0$$

**Proof** (of Lemma 1.16)  Without loss of generality we assume that $r = 1$. Then, with $u = 0$, system (1) has the form

(4)
$$\begin{cases} \dot{x} = Ax + \left(f(x) - A(x)\right) + p(x)q \\ y = h(x) \end{cases}$$

where $A = \dfrac{\partial f}{\partial x}(0)$. Since $\sigma(A) \subset \mathbb{C}^-$ there exist positive constants $M_0 \geq 1$ and $\lambda$ such that $\|e^{At}\| \leq M_0 e^{-\lambda t}$ for all $t$ ($\|\ \|$ denotes matrix sup-norm).

Moreover, there exists a positive constant $M$ such that $|p(x)| \leq M$ for all $x \in \mathbb{R}^n$.

Let $B_{0,r}$ denote the open ball with center $x = 0$ and radius $r$. Choose an arbitrary positive constant $\epsilon$ for which $\lambda - \epsilon M_0 > 0$ and let $r_0$ be such that if $x \in B_{0,r_0}$ and $x \neq 0$, then $|f(x) - Ax| < \epsilon|x|$. Finally, let $0 < \delta_0 < r_0$ be arbitrary.

Define $\mathcal{O} := B_{0,r_0}$, $\tilde{\mathcal{O}} := B_{0,\delta_0/M_0}$ and let $D$ be an arbitrary positive constant such that

$$(5) \qquad D < \frac{r_0-\delta_0}{M_0 M} (\lambda-\epsilon M_0)$$

Suppose that $x_0 \in \tilde{\mathcal{O}}$ and $|q(t)| \leq D$ for all $t$. By continuity, it is obvious that $x(t) \in \mathcal{O}$ for small $t$. We will show by contradiction that $x(t) \in \mathcal{O}$ for all $t \geq 0$.

Assume that there exists a positive $t_0$ such that $x(t_0) \notin \mathcal{O}$. Then there exists a positive constant $T$ such that $x(t) \in \mathcal{O}$ for $0 \leq t < T$ and $|x(T)| = r_0$.

It follows from

$$(6) \qquad x(t) = e^{At}x_0 + \int_0^t e^{A(t-s)}\big(f(x(s)) - Ax(s) + p(x(s))q(s)\big)ds$$

that for $0 \leq t \leq T$

$$(7) \qquad |x(t)| \leq \|e^{At}\|\,|x_0| + \int_0^t \|e^{A(t-s)}\|\big(|f(x(s))-Ax(s)| + |p(x(s))|\,|q(s)|\big)ds$$

$$\leq M_0 e^{-\lambda t}|x_0| + \int_0^t M_0 e^{-\lambda(t-s)}\big(\epsilon|x(s)| + MD\big)ds,$$

and, consequently,

$$(8) \qquad e^{\lambda t}|x(t)| \leq M_0|x_0| + \epsilon M_0 \int_0^t e^{\lambda s}|x(s)|ds + M_0 MD \frac{1}{\lambda}(e^{\lambda t}-1)$$

Now application of Lemma A1 to (8) results in

$$(9) \qquad e^{\lambda t}|x(t)| \leq M_0|x_0| + M_0 MD \frac{1}{\lambda}(e^{\lambda t}-1) +$$

$$+ \epsilon M_0 \int_0^t e^{\epsilon M_0(t-s)}\big(M_0|x_0| + M_0 MD \frac{1}{\lambda}(e^{\lambda s}-1)\big)ds$$

$$= M_0|x_0|e^{\epsilon M_0 t} + M_0 MD \frac{1}{\lambda}(e^{\lambda t}-e^{\epsilon M_0 t}) + \epsilon M_0^2 MD \frac{1}{\lambda}\frac{1}{\lambda-\epsilon M_0}(e^{\lambda t}-e^{\epsilon M_0 t})$$

$$= M_0 \left| x_0 \right| e^{\epsilon M_0 t} + M_0 MD \; \frac{1}{\lambda - \epsilon M_0} \; (e^{\lambda t} - e^{\epsilon M_0 t}), \quad 0 \le t \le T$$

Finally, we have

$$(10) \qquad \left| x(t) \right| \le M_0 \left| x_0 \right| e^{(\epsilon M_0 - \lambda)t} + M_0 MD \; \frac{1}{\lambda - \epsilon M_0} \; (1 - e^{(\epsilon M_0 - \lambda)t})$$

$$\le M_0 \left| x_0 \right| + M_0 MD \; \frac{1}{\lambda - \epsilon M_0} \le \delta_0 + M_0 MD \; \frac{1}{\lambda - \epsilon M_0} < r_0, \quad 0 \le t \le T$$

In particular, $\left| x(T) \right| < r_0$. This contradicts the assumption $\left| x(T) \right| = r_0$. Hence, $x(t) \in \mathcal{O}$ for all $t \ge 0$.

We conclude that if $\left| x_0 \right|$ and $D$ are small enough, then $x(t)$ is bounded.   □

In this appendix we prove Proposition 6.2.5. For convenience we recall the main results from Section 6.2 and Proposition 6.2.5 here.
Consider the square analytic system

$$
(1) \qquad
\begin{cases}
\dot{x} = f(x) + g(x)u, & x \in \mathbb{R}^n, \ u \in \mathbb{R}^m \\[2mm]
y = h(x), & y \in \mathbb{R}^m
\end{cases}
$$

Then Singh's algorithm starts as follows:

### Step 1

Calculate

$$
(2) \qquad \dot{y} = \frac{\partial h}{\partial x}\bigl(f(x) + g(x)u\bigr) =: a_1(x) + b_1(x)u
$$

and define

$$
(3) \qquad \rho_1 := s_1 := \operatorname{rank} b_1(x)
$$

Permute, if necessary, the outputs in such a way that the first $\rho_1$ rows of $b_1(x)$ are linearly independent. Denote the first $\rho_1$ rows of $y$ by $\tilde{y}_1 = (\tilde{y}_{1,1}, \ldots, \tilde{y}_{1,\rho_1})^T$, and the others by $\bar{y}_1$. Since the last rows of $b_1(x)$ are linearly dependent on the first $\rho_1$ rows, the equations (2) can be rewritten as

$$
(4a) \qquad \dot{\tilde{y}}_1 = \tilde{a}_1(x) + \tilde{b}_1(x)u
$$

$$
(4b) \qquad \dot{\bar{y}}_1 = \dot{\bar{y}}_1(x, \dot{\tilde{y}}_1)
$$

where the last equations are affine in $\dot{\tilde{y}}_1$.

### Step 2

Calculate

$$
(5) \qquad \bar{y}_1^{(2)} = \frac{\partial}{\partial x}\,\bar{y}_1^{(1)}\bigl(f(x) + g(x)u\bigr) + \sum_{i=1}^{\rho_1} \frac{\partial \bar{y}_1^{(1)}}{\partial \tilde{y}_{1,i}^{(1)}}\,\tilde{y}_{1,i}^{(2)} =
$$

$$
\qquad = a_2(x, \tilde{y}_1^{(1)}, \tilde{y}_1^{(2)}) + b_2(x, \tilde{y}_1^{(1)})u
$$

Define

(6)    $B_2 := \left(\tilde{b}_1^T, b_2^T\right)^T$  ,   $\rho_2 := \text{rank } B_2(x, \tilde{y}_1^{(1)})$

Permute, if necessary, the components of $\bar{y}_1^{(2)}$ in such a way that the first $\rho_2$ rows of $B_2$ are linearly independent. Let $\tilde{y}_2^{(2)}$ denote the first $s_2 := (\rho_2 - \rho_1)$ rows of $\bar{y}_1^{(2)}$ and $\bar{y}_2^{(2)}$ the remaining rows. Since the last rows of $B_2$ are linearly dependent on the first $\rho_2$ rows, we can write the equations (4,5) as

(7)    $$\begin{cases} \dot{\tilde{y}}_1 &= \tilde{a}_1(x) + \tilde{b}_1(x)u \\[2mm] \dot{\tilde{y}}_2^{(2)} &= a_2(x, \tilde{y}_1^{(1)}, \tilde{y}_1^{(2)}) + \tilde{b}_2(x, \tilde{y}_1^{(1)})u \\[2mm] \dot{\bar{y}}_2^{(2)} &= \bar{y}_2^{(2)}(x, \tilde{y}_1^{(1)}, \tilde{y}_1^{(2)}, \tilde{y}_2^{(2)}) \end{cases}$$

The general step of Singh's algorithm can be found in Section 6.2. Assume in the sequel that the system (1) is strongly dynamically input-output decouplable and that $x_0$ is a strongly regular point for Singh's algorithm. Suppose that Singh's algorithm stops after $\alpha$ steps. Then we have

(8)    $\dot{\tilde{Y}}_\alpha = \tilde{A}_\alpha\left(x, \{\tilde{y}_i^{(j)} \mid 1 \le i \le \alpha-1, \ i \le j \le \alpha\}\right) +$

   $+ \tilde{B}_\alpha\left(x, \{\tilde{y}_i^{(j)} \mid 1 \le i \le \alpha-1, \ i \le j \le \alpha-1\}\right)u$

where $\tilde{Y}_\alpha = \left(\tilde{y}_1^T, \ldots, \tilde{y}_\alpha^{(\alpha-1)T}\right)^T$ and $\tilde{B}_\alpha\left(x, \{\tilde{y}_i^{(j)} \mid 1 \le i \le \alpha-1, \ i \le j \le \alpha-1\}\right)$ has full rank $m$. From (8) we obtain

(9)    $u = \tilde{B}_\alpha^{-1}\left(\dot{\tilde{Y}}_\alpha - \tilde{A}_\alpha\right) =: \varphi\left(x, \{\tilde{y}_i^{(j)} \mid 1 \le i \le \alpha, \ i \le j \le \alpha\}\right)$

Let $\gamma_i$ be the lowest time-derivative and $\delta_i$ the highest time-derivative of $\tilde{y}_i$ ($i = 1, \ldots, m$) appearing in (9). Then it can quite easily be shown that $\varphi$ is of the form

(10)    $\varphi\left(x, \{\tilde{y}_i^{(j)} \mid 1 \le i \le m, \ \gamma_i \le j \le \delta_i\}\right)$

   $= \varphi_1\left(x, \{\tilde{y}_i^{(j)} \mid 1 \le i \le m, \ \gamma_i \le j \le \delta_i - 1\}\right)$

   $+ \sum_{i=1}^{m} \varphi_{2i}\left(x, \{\tilde{y}_i^{(j)} \mid 1 \le i \le m, \ \gamma_i \le j \le \delta_i - 1\}\right)\tilde{y}_i^{(\delta_i)}$

Let $z_i$ ($i = 1, \ldots, m$) be a vector of dimension $\delta_i - \gamma_i$, then the system

$$(11) \quad \begin{cases} \dot{z}_i = A_i z_i + B_i v_i, \quad i = 1,\ldots,m \\[2mm] u = \varphi_1(x,z_1,\ldots,z_m) + \sum_{i=1}^{m} \varphi_{2i}(x,z_1,\ldots,z_m)v_i \end{cases}$$

with inputs $v$, outputs $u$ and

$$(12) \quad A_i = \begin{bmatrix} 0 & 1 & & 0 \\ & \ddots & \ddots & \\ & & & 1 \\ 0 & & & 0 \end{bmatrix}_{(\delta_i-\gamma_i)\times(\delta_i-\gamma_i)} \qquad B_i = \begin{bmatrix} 0 \\ \vdots \\ 0 \\ 1 \end{bmatrix}_{\delta_i-\gamma_i} , \quad i = 1,\ldots,m$$

solves the SLDIODP for (1).

**Proposition 6.2.5** Consider the square analytic system (1). Assume that $x_0$ is a strongly regular point for Singh's algorithm. Then the constants $\gamma_i$ and $\delta_i$ ($i = 1,\ldots,m$) are intrinsic. Suppose that (11,12) and

$$(13) \quad \begin{cases} \dot{\bar{z}}_i = \bar{A}_i \bar{z}_i + \bar{B}_i v_i , \quad i = 1,\ldots,m \\[2mm] u = \bar{\varphi}_1(x,\bar{z}_1,\ldots,\bar{z}_m) + \sum_{i=1}^{m} \bar{\varphi}_{2i}(x,\bar{z}_1,\ldots,\bar{z}_m)v_i \end{cases}$$

with

$$(14) \quad \bar{A}_i = \begin{bmatrix} 0 & 1 & & 0 \\ & \ddots & \ddots & \\ & & & 1 \\ 0 & & & 0 \end{bmatrix}_{(\bar{\delta}_i-\bar{\gamma}_i)\times(\bar{\delta}_i-\bar{\gamma}_i)} \qquad \bar{B}_i = \begin{bmatrix} 0 \\ \vdots \\ 0 \\ 1 \end{bmatrix}_{\bar{\delta}_i-\bar{\gamma}_i} , \quad i = 1,\ldots,m$$

denote two arbitrary Singh compensators obtained for the system (1), then there exist initial points $z_0$ and $\bar{z}_0$ and an analytic coordinate transformation $(x,\bar{z}) = \Phi(x,z)$ defined locally around $(x_0,z_0)$ transforming the compensator (11,12) in (13,14).

Recall that a Singh compensator is a regular dynamic state feedback of the form (13,14) obtained by an arbitrary application of Singh's algorithm. The nonuniqueness of the Singh compensator is caused by the freedom at each step of the algorithm to permute the outputs in such a way that the first $\rho_k$ rows of $B_k$ are linearly independent. The fact that the constants $\gamma_i$ and $\delta_i$ are intrinsic means that $\{(\gamma_1,\delta_1),\ldots,(\gamma_m,\delta_m)\}$ is an invariant set for the system (1), i.e. the set does not depend on the permutation of the outputs at each step of the algorithm. Since only a finite number of permutations is possible, the general result in Proposition 6.2.5 follows from checking what happens if in the first step of the algorithm one output in $\tilde{y}_1$ is interchanged with one of the outputs in $\bar{y}_1$. It follows from the proof that the compensators (11,12), (13,14) can be calculated step by step and that there exists a one-to-one correspondence between $\bar{z}_i$, $i = 1,\ldots,m$ and $z_i$, $i = 1,\ldots,m$. This relation establishes the diffeomorphism between $(x,\bar{z})$ and $(x,z)$.

**Remark** It is shown in [HNvdW4] that every Singh compensator has the same dimension $\sum_{i=1}^{m} (\delta_i - \gamma_i)$ and that the constants $\delta_i$ are equal to the essential orders $\epsilon_i$ (cf. [GM]). This result immediately implies that the constants $\delta_i$ are intrinsic, because the essential orders are (see [Mo]). □

**Proof** (of Proposition 6.2.5) We assume, without loss of generality, that there is choice in the first step of the algorithm and that the outputs $y_{\rho_1}$ and $y_{\rho_1+1}$ may be interchanged. So we assume that

$$(15) \qquad \text{rank } \tilde{B}_1(x) = \text{rank} \begin{bmatrix} b_{1,1}(x) \\ \vdots \\ b_{1,\rho_1-1}(x) \\ b_{1,\rho_1}(x) \end{bmatrix} = \text{rank} \begin{bmatrix} b_{1,1}(x) \\ \vdots \\ b_{1,\rho_1-1}(x) \\ b_{1,\rho_1+1}(x) \end{bmatrix} \quad (=\rho_1)$$

Then there exists a row vector $\Gamma_1(x) = (\Gamma_{1,1}(x),\ldots,\Gamma_{1,\rho_1-1}(x))$ and a nonzero function $\Gamma_{1,\rho_1}(x)$ such that, with $\tilde{b}_{11}(x)=(b_{1,1}(x),\ldots,b_{1,\rho_1-1}(x))^T$,

$$(16)$$

$$b_{1,\rho_1+1}(x) = \Gamma_{1,1}(x)b_{1,1}(x) + \ldots + \Gamma_{1,\rho_1-1}(x)b_{1,\rho_1-1}(x) + \Gamma_{1,\rho_1}(x)b_{1,\rho_1}(x)$$

$$= \Gamma_1(x)\tilde{b}_{11}(x) + \Gamma_{1,\rho_1}(x)b_{1,\rho_1}(x)$$

It follows from (16) that

$$(17) \qquad \begin{bmatrix} \tilde{b}_{11}(x) \\ b_{1,\rho_1+1}(x) \end{bmatrix} = \begin{bmatrix} I_{\rho_1-1} & 0 \\ \Gamma_1(x) & \Gamma_{1,\rho_1}(x) \end{bmatrix} \begin{bmatrix} \tilde{b}_{11}(x) \\ b_{1,\rho_1}(x) \end{bmatrix} =: \tilde{\Gamma}_1(x)\tilde{b}_1(x)$$

In the sequel we refer to the first application of Singh's algorithm if the outputs $\tilde{y}_{1,1},\ldots,\tilde{y}_{1,\rho_1}$ are taken in the first step of the algorithm (see the equations (4)-(7)) and to the second application if we interchange the outputs $y_{\rho_1}$ and $y_{\rho_1+1}$ in the first step of the algorithm. We use the superscript ' to refer to the second application of Singh's algorithm. So, $\tilde{y}_1 = (\tilde{y}_{11}^T, y_{\rho_1})^T$ with $\tilde{y}_{11} = (\tilde{y}_{1,1}, \ldots, \tilde{y}_{1,\rho_1-1})^T$ and $\tilde{y}_1' = (\tilde{y}_{11}^T, y_{\rho_1+1})^T$. Let $\tilde{b}_1^+(x)$ be a right-inverse of $\tilde{b}_1(x)$, then it follows from (4a) that

$$(18) \qquad u = \tilde{b}_1^+(x) \begin{bmatrix} \dot{\tilde{y}}_{11} - \tilde{a}_{11}(x) \\ \dot{y}_{\rho_1} - a_{1,\rho_1}(x) \end{bmatrix}$$

where $\tilde{a}_{11}(x) = (a_{1,1}(x), \ldots, a_{1,\rho_1-1}(x))^T$. The equations (4b) follow from substitution of (18) in

$$(19) \qquad \dot{\tilde{y}}_1 = \begin{bmatrix} \dot{y}_{\rho_1+1} \\ \dot{\tilde{y}}_{11} \end{bmatrix} = \begin{bmatrix} a_{1,\rho_1+1}(x) + b_{1,\rho_1+1}(x)u \\ \bar{a}_{11}(x) + \bar{b}_{11}(x)u \end{bmatrix}$$

using (17):

$$(20a) \qquad \dot{y}_{\rho_1+1} = a_{1,\rho_1+1}(x) + b_{1,\rho_1+1}(x)u$$

$$= a_{1,\rho_1+1}(x) + \left(\Gamma_1(x) \ \Gamma_{1,\rho_1}(x)\right)\tilde{b}_1(x)\tilde{b}_1^+(x) \begin{bmatrix} \dot{\tilde{y}}_{11}-\tilde{a}_{11}(x) \\ \dot{y}_{\rho_1}-a_{1,\rho_1}(x) \end{bmatrix}$$

$$= a_{1,\rho_1+1}(x) + \Gamma_1(x)(\dot{\tilde{y}}_{11}-\tilde{a}_{11}(x)) + \Gamma_{1,\rho_1}(x)(\dot{y}_{\rho_1}-a_{1,\rho_1}(x))$$

and

$$(20b) \qquad \dot{\tilde{y}}_{11} = \bar{a}_{11}(x) + \bar{b}_{11}(x)\tilde{b}_1^+(x) \begin{bmatrix} \dot{\tilde{y}}_{11}-\tilde{a}_{11}(x) \\ \dot{y}_{\rho_1}-a_{1,\rho_1}(x) \end{bmatrix}$$

Differentiation of (20a) and (20b) yields in the second step of the algorithm (cf. (5))

$$(21a) \qquad y_{\rho_1+1}^{(2)} = c(x,\dot{\tilde{y}}_{11},\dot{y}_{\rho_1}) + d(x,\dot{\tilde{y}}_{11},\dot{y}_{\rho_1})u + \left(\Gamma_1(x) \ \Gamma_{1,\rho_1}(x)\right) \begin{bmatrix} \tilde{y}_{11}^{(2)} \\ y_{\rho_1}^{(2)} \end{bmatrix}$$

$$(21b) \qquad \bar{y}_{11}^{(2)} = P(x,\dot{\tilde{y}}_{11},\dot{y}_{\rho_1}) + Q(x,\dot{\tilde{y}}_{11},\dot{y}_{\rho_1})u + \bar{b}_{11}(x)\tilde{b}_1^+(x) \begin{bmatrix} \tilde{y}_{11}^{(2)} \\ y_{\rho_1}^{(2)} \end{bmatrix}$$

Note that it follows from (20a) that

$$(22) \qquad \dot{y}_{\rho_1} = a_{1,\rho_1}(x) + \Gamma_{1,\rho_1}^{-1}(x)\left[\dot{y}_{\rho_1+1}-a_{1,\rho_1+1}(x)-\Gamma_1(x)(\dot{\tilde{y}}_{11}-\tilde{a}_{11}(x))\right]$$

The equations obtained in the second step of the second application of Singh's algorithm immediately follow from (21). Equation (21a) yields

$$(23a) \qquad y_{\rho_1}^{(2)} = \Gamma_{1,\rho_1}^{-1}(x)\left[y_{\rho_1+1}^{(2)} - c(x,\tilde{\bar{y}}_{11},\dot{\bar{y}}_{\rho_1}) - d(x,\tilde{\bar{y}}_{11},\dot{\bar{y}}_{\rho_1})u - \Gamma_1(x)\tilde{y}_{11}^{(2)}\right]$$

$$=: \Gamma_{1,\rho_1}^{-1}(x)\left[-c'(x,\tilde{\bar{y}}_{11},\dot{\bar{y}}_{\rho_1+1}) - d'(x,\tilde{\bar{y}}_{11},\dot{\bar{y}}_{\rho_1+1})u - \Gamma_1(x)\tilde{y}_{11}^{(2)} + y_{\rho_1+1}^{(2)}\right]$$

where $c'$ is obtained by substitution of (22) in $c$, and (21b) and (23a) give

$$(23b) \qquad \tilde{\bar{y}}_{11}^{(2)} = P'(x,\tilde{\bar{y}}_{11},\dot{\bar{y}}_{\rho_1+1}) + Q'(x,\tilde{\bar{y}}_{11},\dot{\bar{y}}_{\rho_1+1}) +$$

$$+ b_{11}(x)\tilde{b}_1^+(x)\left[\begin{array}{c} \tilde{\bar{y}}_{11}^{(2)} \\[2mm] \Gamma_{1,\rho_1}^{-1}(x)\left[-c'(x,\tilde{\bar{y}}_{11},\dot{\bar{y}}_{\rho_1+1}) - d'(x,\tilde{\bar{y}}_{11},\dot{\bar{y}}_{\rho_1+1})u - \Gamma_1(x)\tilde{y}_{11}^{(2)} + y_{\rho_1+1}^{(2)}\right] \end{array}\right]$$

$$= P''(x,\tilde{\bar{y}}_{11},\dot{\bar{y}}_{\rho_1+1}) + Q''(x,\tilde{\bar{y}}_{11},\dot{\bar{y}}_{\rho_1+1})u +$$

$$+ \bar{b}_{11}(x)\tilde{b}_1^+(x)\left[\begin{array}{cc} I_{\rho_1-1} & 0 \\[2mm] -\Gamma_{1,\rho_1}^{-1}(x)\Gamma_1(x) & \Gamma_{1,\rho_1}^{-1}(x) \end{array}\right]\left[\begin{array}{c} \tilde{y}_{11}^{(2)} \\[2mm] y_{\rho_1+1}^{(2)} \end{array}\right]$$

where

(24)

$$Q''(x,\tilde{\bar{y}}_{11},\dot{\bar{y}}_{\rho_1+1}) = Q'(x,\tilde{\bar{y}}_{11},\dot{\bar{y}}_{\rho_1+1}) - \left(\bar{b}_{11}(x)\tilde{b}_1^+(x)\right)_{\rho_1}\Gamma_{1,\rho_1}^{-1}(x)d'(x,\tilde{\bar{y}}_{11},\dot{\bar{y}}_{\rho_1+1})$$

Since the integers $\rho_1, \rho_2,\ldots$ are intrinsic (see [Mo]) it follows that $\rho_2 = \text{rank } B_2 - \text{rank } B_2'$, where

$$(25) \qquad B_2 = \left[\begin{array}{c} \tilde{b}_{11}(x) \\[2mm] b_{1,\rho_1}(x) \\[2mm] d(x,\tilde{\bar{y}}_{11},\dot{\bar{y}}_{\rho_1}) \\[2mm] Q(x,\tilde{\bar{y}}_{11},\dot{\bar{y}}_{\rho_1}) \end{array}\right], \quad B_2' = \left[\begin{array}{c} \tilde{b}_{11}(x) \\[2mm] b_{1,\rho_1+1}(x) \\[2mm] -\Gamma_{1,\rho_1}^{-1}(x)d'(x,\tilde{\bar{y}}_{11},\dot{\bar{y}}_{\rho_1+1}) \\[2mm] Q''(x,\tilde{\bar{y}}_{11},\dot{\bar{y}}_{\rho_1+1}) \end{array}\right]$$

Since $d'$ and $Q'$ are obtained by substituting (22) in $d$ and $Q$, it is clear from (24) that the rows that contribute to the rank of $B_2$ are the same as the rows that contribute to the rank of $B_2'$.

The following possibilities may occur (for the first application of Singh's algorithm):

(i) $d(x,\tilde{y}_{11},\dot{y}_{\rho_1})$ is linearly dependent on $\tilde{b}_1(x)$;

(ii) $d(x,\tilde{y}_1,\dot{y}_{\rho_1})$ is independent of $\tilde{b}_1(x)$, but $y_{\rho_1+1}^{(2)}$ is not chosen in $\tilde{y}_2^{(2)}$;

(iii) $y_{\rho_1+1}^{(2)}$ is chosen in $\tilde{y}_2^{(2)}$.

First, consider case (iii). It follows from the foregoing that in the second application of Singh's algorithm $y_{\rho_1}^{(2)}$ may be chosen in $\tilde{y}_2^{(2)}$.
So we have for the first application of Singh's algorithm

(26)
$$
\left\{
\begin{aligned}
\dot{\tilde{y}}_{11} &= \tilde{a}_{11}(x) + \tilde{b}_{11}(x)u \\
\dot{y}_{\rho_1} &= a_{1,\rho_1}(x) + b_{1,\rho_1}(x)u \\
y_{\rho_1+1}^{(2)} &= c(x,\tilde{y}_{11},\dot{y}_{\rho_1}) + d(x,\tilde{y}_{11},\dot{y}_{\rho_1})u + \Gamma_1(x)\tilde{y}_{11}^{(2)} + \Gamma_{1,\rho_1}(x)y_{\rho_1}^{(2)} \\
\tilde{y}_{22}^{(2)} &= \tilde{P}(x,\tilde{y}_{11},\dot{y}_{\rho_1}) + \tilde{Q}(x,\tilde{y}_{11},\dot{y}_{\rho_1})u + \tilde{B}(x)\begin{bmatrix} \tilde{y}_{11}^{(2)} \\ y_{\rho_1}^{(2)} \end{bmatrix}
\end{aligned}
\right.
$$

where $\tilde{y}_2 = (y_{\rho_1+1},\tilde{y}_{22}^T)^T$ and $\tilde{P}$ denotes the first $\rho_2-(\rho_1+1)$ rows of $P$ etc.

For the second application we have (cf. (23))

(27)
$$
\left\{
\begin{aligned}
\dot{\tilde{y}}_{11} &= a_{11}(x) + \tilde{b}_{11}(x)u \\
\dot{y}_{\rho_1+1} &= a_{1,\rho_1+1}(x) + b_{1,\rho_1+1}(x)u \\
y_{\rho_1}^{(2)} &= \Gamma_{1,\rho_1}^{-1}(x)\left[-c'(x,\tilde{y}_{11},\dot{y}_{\rho_1+1})-d'(x,\tilde{y}_{11},\dot{y}_{\rho_1+1})u-\Gamma_1(x)\tilde{y}_{11}^{(2)}+y_{\rho_1+1}^{(2)}\right] \\
\tilde{y}_{22}^{(2)} &= \tilde{P}''(x,\tilde{y}_{11},\dot{y}_{\rho_1+1}) + \tilde{Q}''(x,\tilde{y}_{11},\dot{y}_{\rho_1+1})u + \tilde{B}(x)\tilde{\Gamma}_1^{-1}(x)\begin{bmatrix} \tilde{y}_{11}^{(2)} \\ y_{\rho_1+1}^{(2)} \end{bmatrix}
\end{aligned}
\right.
$$

It is clear from (26) and (27) that in the first case after two steps $y_{\rho_1}^{(1)}$, $y_{\rho_1}^{(2)}$ and $y_{\rho_1+1}^{(2)}$ appear and in the second case $y_{\rho_1+1}^{(1)}$, $y_{\rho_1+1}^{(2)}$ and $y_{\rho_1}^{(2)}$. The second derivatives of $\tilde{y}_{11}$ appear in (26) and (27) at the same time. This follows from the following arguments. For outputs $\tilde{y}_{1,i}$ for which

$\Gamma_{1,i}(x) \neq 0$ it follows that $\tilde{y}_{1,i}^{(2)}$ appears in both $y_{\rho_1}^{(2)}$ and $y_{\rho_1+1}^{(2)}$. For outputs $\tilde{y}_{1,i}$ with $\Gamma_{1,i}(x) = 0$ we have that the coefficient of $\tilde{y}_{1,i}^{(2)}$ in $\tilde{B}(x)$ and $\tilde{B}(x)\tilde{\Gamma}_1^{-1}(x)$ is the same (viz. $\tilde{B}_i(x)$), so these second derivatives appear depending on $\tilde{B}_i(x)$.

Repeating the above arguments, it can be proved that in the third step of Singh's algorithm the same outputs can be chosen in $\tilde{y}_3^{(3)}$ and that the same derivatives of $\tilde{y}_{11}$, $y_{\rho_1}$, $y_{\rho_1+1}$ and $\tilde{y}_{22}$ appear in both applications of the algorithm. This yields at the $\alpha$-th step that the constants $\gamma_i$ and $\delta_i$ are the same for $i = 1,\ldots,\rho_1-1$, $\rho_1+2,\ldots,m$ for both applications of Singh's algorithm and that in the first case $\gamma_{\rho_1} = 1$, $\gamma_{\rho_1+1} = 2$ and in the second case $\gamma_{\rho_1} = 2$, $\gamma_{\rho_1+1} = 1$ (and $\delta_{\rho_1} = \delta_{\rho_1+1}$).

It can be seen from the algorithm that the construction of the compensator can be accomplished step by step. If $\Gamma_{1,i}(x) = 0$ and $\tilde{B}_i(x) = 0$ then $\tilde{y}_{1,i}^{(2)}$ ($= \tilde{y}_{11,i}^{(2)}$) does not appear in $y_{\rho_1+1}^{(2)}$ or $\bar{y}_{22}^{(2)}$, so no second derivatives of $\tilde{y}_{1,i}$ occur in the second step of the algorithm (but, of course, second or higher derivatives of $\tilde{y}_{1,i}$ may appear in subsequent steps). In all other cases, the second derivative of the output $\tilde{y}_{1,i}$ appears in the second step, so $\tilde{y}_{1,i}$ is differentiated at least once (once if the algorithm stops in 2 steps). So in that case we have ($y_{\rho_1}^{(2)}$ appears in any case!)

$$
(28) \qquad
\begin{cases}
\tilde{y}_{1,i} = z_i & , \; i = 1,\ldots,\rho_1-1, \quad \tilde{B}_i(x) \neq 0 \text{ or } \Gamma_{1,i}(x) \neq 0 \\
y_{\rho_1} = z_{\rho_1}
\end{cases}
$$

If $\tilde{y}_{1,i} = z_i$, $i = 1,\ldots,\rho_1-1$, then it follows from (18) that

$$
(29) \qquad
u = \tilde{b}_1^+(x)
\begin{bmatrix}
z_1 - a_{1,1}(x) \\
\vdots \\
z_{\rho_1-1} - a_{1,\rho_1-1}(x) \\
z_{\rho_1} - a_{1,\rho_1}(x)
\end{bmatrix}
$$

In the second application we have (assuming that $\Gamma_{1,i}(x) \neq 0$, $i=1,\ldots,\rho_1-1$)

$$
(30) \qquad
\begin{cases}
\tilde{y}_{1,i} = \bar{z}_i & , \; i = 1,\ldots,\rho_1-1 \\
y_{\rho_1+1} = \bar{z}_{\rho_1+1}
\end{cases}
$$

and

$$
(31) \qquad u = \begin{bmatrix} \bar{b}_{11}(x) \\ \\ b_{1,\rho_1+1}(x) \end{bmatrix}^+ \begin{bmatrix} \bar{z}_1 - a_{1,1}(x) \\ \vdots \\ \bar{z}_{\rho_1-1} - a_{1,\rho_1-1}(x) \\ \bar{z}_{\rho_1+1} - a_{1,\rho_1+1}(x) \end{bmatrix}
$$

$$
- \bar{b}_1^+(x)\bar{\Gamma}_1^{-1}(x) \begin{bmatrix} \bar{z}_1 - a_{1,1}(x) \\ \vdots \\ \bar{z}_{\rho_1-1} - a_{1,\rho_1-1}(x) \\ \bar{z}_{\rho_1+1} - a_{1,\rho_1+1}(x) \end{bmatrix}
$$

It is clear from (28), (30) and (20a) that

$$
(32) \quad \begin{cases} \bar{z}_i = z_i \quad , \quad i = 1,\ldots,\rho_1-1 \\ \\ \bar{z}_{\rho_1+1} = a_{1,\rho_1}(x) + \sum_{i=1}^{\rho_1-1} \Gamma_{1,i}(x)(z_i - a_{1,i}(x)) + \Gamma_{1,\rho_1}(x)(z_{\rho_1} - a_{1,\rho_1}(x)) \end{cases}
$$

and if we denote the inputs given by (29) and (31) by $u_{(1)}$ and $u_{(2)}$, respectively, it follows easily that $u_{(1)} = u_{(2)}$. The compensators (11,12) and (13,14) can be constructed now step by step. Similar arguments as used above show that there exists a diffeomorphism $(x,\bar{z}) = \Phi(x,z)$ mapping the compensator (11,12) on (13,14).

Now, consider the cases (i) and (ii) (see page 127).
Suppose that

$$
(33) \qquad y_{\rho_1+1}^{(3)} = \sigma_1(x,\tilde{y}_1^{(1)},\tilde{y}_1^{(2)},\tilde{y}_1^{(3)},\tilde{y}_2^{(2)},\tilde{y}_2^{(3)}) + \sigma_2(x,\tilde{y}_1^{(1)},\tilde{y}_1^{(2)},\tilde{y}_2^{(2)})u
$$

and that $\sigma_2$ is not linearly dependent on the rows in $\tilde{B}_2$. Then $y_{\rho_1+1}^{(3)}$ may be chosen in $\tilde{y}_3^{(3)}$ (and in the second application $y_{\rho_1}^{(3)}$). So now $\gamma_{\rho_1} = 1$, $\gamma_{\rho_1+1} = 3$ ($\gamma_{\rho_1} = 3$, $\gamma_{\rho_1+1} = 1$). The other arguments used above are still valid with some minor modifications. Also in this case there exists a one-to-one correspondence between the compensators (11,12) and (13,14). Note that in this case we also have equation (28) as a first step in the construction of the compensator. □

# LIST OF ABBREVIATIONS

Below some frequently used subspaces and distributions are listed.

$\mathcal{V}^*$ largest controlled invariant subspace in the kernel of the output mapping

$\mathcal{V}_s^*$ largest stabilizability subspace in the kernel of the output mapping

$\mathcal{R}^*$ largest controllability subspace in the kernel of the output mapping

$\Delta^*$ largest locally controlled invariant distribution in the kernel of the output mapping

$\Pi^*$ largest local controllability distribution in the kernel of the output mapping

$\Delta_s^*$ largest stabilizability distribution in the kernel of the output mapping

$(\Delta^P)_*$ smallest constant dimensional locally controlled invariant distribution in the kernel of the output mapping that contains $\Pi^*$ and the disturbance vector fields $P$

## Subject Index

# Lecture Notes in Control and Information Sciences

Edited by M. Thoma and A. Wyner

# Lecture Notes in Control and Information Sciences

Edited by M. Thoma and A. Wyner

# Lecture Notes in Control and Information Sciences

Edited by M. Thoma and A. Wyner